MODERN TOOLS FOR GENETIC ENGINEERING

Edited by **Michael S.D. Kormann**

Modern Tools for Genetic Engineering

http://dx.doi.org/10.5772/61377
Edited by Michael S.D. Kormann

Contributors

Lei Yu, Biao Lu, Jennifer Batara, Florence Rouleux-Bonnin, Colette Bastie, Francisco Martin, Guan-Huei Lee, Myo Myint Aung, Jose C Segovia, Oscar Quintana-Bustamante, Sergio López-Manzaneda, Susana Navarro, Maria Garcia-Bravo, Virginia Nieto-Romero, Sara Fañanás-Baquero, Francisco Jose Roman-Rodriguez, Maria Fernandez-Garcia, María José Pino-Barrio, Fatima Rodriguez-Fornes, Begoña Diez-Cabezas, Houria Bachtarzi, Tim Farries, Stefanie Krajewski, Tatjana Michel, Hans Peter Wendel, Andrew Handorf, Hans Sollinger, Tausif Alam

CBS, Edition **2017**

Published by InTech

Janeza Trdine 9, 51000 Rijeka, Croatia

Publishing Process Manager Sandra Bakic

Technical Editor InTech DTP team

Cover Designer InTech Design Team

Additional hard copies can be obtained from orders@intechopen.com

Modern Tools for Genetic Engineering, Edited by Michael S.D. Kormann
p. cm.
Print ISBN 978-953-51-2392-7
Online ISBN 978-953-51-2234-0

INTECH

open science | open minds

Contents

Contents

Preface

Treatment of diseases caused by dysfunctional or insufficiently expressed proteins can principally be addressed by either administration of the functionally active protein itself or by its genetically encoded precursors, the corresponding gene (DNA) or its transcript (mRNA). Administration of the functional protein has been utilized for the successful treatment of a number of diseases, including hemophilia and alpha-1 antitrypsin deficiency. However, mode of delivery remains a major obstacle to protein therapy, especially in non-systemic diseases, where administration via standard routes, such as oral, intravenous (i.v.) or intramuscular (i.m.) injection, is often ineffective, as the therapeutic protein is metabolized or cleared before it can enter the target tissue or target cells.

A promising alternative to gene addition and transcript replacement approaches is gene correction through editing of the endogenous gene. In this approach, a mutation is corrected in its' endogenous chromosomal locus using specific endonucleases such as Zinc Finger Nucleases (ZFNs), TAL Effector Nucleases (TALENs) or CRISPR/Cas9 systems, together with a donor template. The endonuclease cuts surgically at the predetermined locus in the genome, creating a double-strand break, which initiates homologous recombination. During the recombination a manipulation or correction template will be favored over the homologous chromosome if present in excess. Correcting a gene defect on the genomic level, issues such as inappropriate tissue specificity, timing, level and duration of expression can be completely avoided, because the targeted gene remains under natural, endogenous controls.

Gene therapy technology today is more effective and precise as it has ever been. The progress of tools and its application is highlighted in this book.

Enjoy the fascinating insight in modern genetic engineering.

Michael S.D. Kormann, PhD
University Children's Hospital - Section I,
Pediatric Infectiology & Immunology,
Translational Genomics and Gene Therapy,
Lothar-Meyer-Bau,
Tübingen, Germany

Section 1

Gene Delivery Technologies

Chapter 1

Next-Generation Therapeutics: mRNA as a Novel Therapeutic Option for Single-Gene Disorders

Tatjana Michel, Hans-Peter Wendel and
Stefanie Krajewski

Additional information is available at the end of the chapter

http://dx.doi.org/10.5772/62243

Abstract

In single-gene disorders, such as α1-antitrypsin deficiency (AATD), hemophilia B (clotting factor IX deficiency), and lecithin-cholesterol acyltransferase deficiency (LCATD), a gene mutation causes missing or dysfunctional protein synthesis, which in turn can lead to serious complications for the patient affected. Furthermore, single-gene disorders are associated with severe, early-onset conditions and necessitate expensive lifelong care. Today, therapeutic treatment options remain limited, cost-intensive, or ineffective. Therefore, the novel mRNA-based therapeutic strategy for the treatment of single-gene disorders, which is based on the induction of *de novo* synthesis of the functional proteins, has extraordinary potential. After the delivery of the specific mRNA to the target cells, the desired protein is expressed by the cells' own translational machinery, and hence, a fully functional protein replaces the defective or missing protein. mRNA therapy provides an innovative, highly promising, and inexpensive therapeutic approach and will thus lead to new advances in the treatment of single-gene disorders.

Keywords: α1-antitrypsin deficiency, hemophilia B, lecithin-cholesterol acyltransferase deficiency, mRNA therapy, single-gene disorders, messenger RNA, next-generation therapeutics, gene mutation

1. Introduction

The human body is made up of millions of cells, which have special, well-defined functions, such as the transportation of oxygen in the blood. Proteins carry out all these functions that are necessary for life. In most of the cells, the genetic information is encoded on the 23 pairs of chromosomes found in the nucleus. On each chromosome, the information for the production of a great range of different proteins is contained in genes made up of DNA. There are approximately 25,000 protein-encoding genes in the human genome [1].

Protein synthesis takes place in two major processes. First, the DNA is transcribed into the mRNA in the nucleus followed by cytoplasmic mRNA translation into the protein.

Alteration in a single gene (i.e., a single-gene disorder) caused by a mutation in the gene's DNA sequences leads to dysfunction of the gene. As a result, the protein the gene codes for is either altered or missing, which can result in serious complications in the human body.

Over the last few years, it has become apparent that single-gene disorders are far more numerous than previously assumed, and more than 1800 single-gene disorders have been identified [2]. Moreover, single-gene disorders are associated with severe, early-onset conditions necessitating lifelong care [3].

2. Examples of single-gene disorders

2.1. α1-Antitrypsin (AAT) deficiency (AATD)

AAT is a serine protease inhibitor belonging to the serpin superfamily, and it is predominantly synthesized in the liver and released into the bloodstream. By inhibiting neutrophil elastase, which is released from activated neutrophil granulocytes and macrophages, AAT plays a pivotal role in the prevention of proteolytic damage to the host tissue in the presence of inflammatory processes. Physiological AAT plasma concentrations are within the range of 20 to 53 μmol/L (150–300 mg/dL).

Mutations in the SERPIN1A gene are associated with AATD, one of the most common metabolic diseases in Europe. AATD, first described in 1963 by Laurell and Erikson, affects approximately 1 in 2000 to 1 in 5000 individuals [4, 5]. People suffering from AATD have AAT levels below 11 μmol/L (80 mg/dL) and are predisposed to severe lung diseases, such as chronic obstructive pulmonary disease (COPD), or liver diseases.

2.2. Hemophilia B

Hemophilia B, also called Christmas disease, is the second most common form of hemophilia, affecting approximately 20% of those diagnosed with hemophilia. It is a bleeding disorder that is typically inherited and characterized by a lack of clotting activity of factor IX (FIX; Christmas factor), a blood clotting factor that is synthesized in hepatocytes and plays a crucial role in blood coagulation. Depending on the bleeding phenotype of hemophilia B, which is classified as mild, moderate, or severe, and the sufferer's overall health, symptoms vary from prolonged, partially excessive bleeding to serious bruising and joint pain [6].

2.3. Lecithin-cholesterol acyltransferase (LCAT) deficiency (LCATD)

The gene for LCAT is localized on chromosome 16 and primarily expressed in the liver. After secretion to the plasma, it is primarily found attached to circulating high-density lipoprotein (HDL) particles [7]. LCAT converts free cholesterol into cholesterol esters on the surface of HDL, thus removing cholesterol from the blood and tissues [8].

Mutations in the LCAT gene result in deficient or absent catalytic LCAT activity, leading to a reduction in the enzyme's ability to attach cholesterol to lipoproteins. Hence, deficiency leads to the accumulation of unesterified cholesterol in different tissues (e.g., in the cornea, erythrocytes, or kidneys) and may lead to corneal opacities, renal failure, or hemolytic anemia [9]. Due to the accumulation of cholesterol in the lining of the arteries, LCATD sufferers have an increased risk for premature atherosclerosis.

2.4. Familial hypercholesterolemia (FH)

FH is a genetic disease characterized by high low-density lipoprotein (LDL) cholesterol levels in the blood. This is caused by a defect in the gene for the LDL receptor (LDLR) that prevents it from absorbing the LDL from the blood into the cell for metabolization. The symptoms in patients with FH range from harmless fatty skin deposits called xanthoma to life-threatening atherosclerotic vascular disease, which can culminate in myocardial infarction or stroke [10]. In patients suffering from the homozygous form of FH, such life-threatening complications occur in infancy. Statins are currently used for standard therapy, but their efficacy is controversial [11].

2.5. Available treatment options

Different treatment options exist depending on the genetic disorder. Treatment of AATD is currently achieved by aerosol or intravenous augmentation therapy of purified and pooled human plasma AAT protein [12]. Furthermore, the risks of proinflammatory stimuli to the lung need to be minimized by ensuring that the patient abstains from smoking, using bronchodilators, etc. Augmentation therapy is, however, very expensive, because the AAT protein derived from a healthy donor needs to undergo a rigorous screening process before it is ready to be infused [13].

Patients suffering from hemophilia B are routinely treated with recombinant FIX concentrates, which have greatly reduced the mortality associated with hemophilia B. However, there are still significant drawbacks of this existing therapy, including the necessity of multiple weekly infusions for patients with severe hemophilia B as well as repeated bleeding despite prophylactic therapy, which can cause long-term damage in joints and other tissues.

The therapeutic up-regulation of LCAT function has gained interest in recent years, not only as an enzyme replacement therapy for LCATD syndromes but also as a potential therapeutic strategy for reducing atherosclerosis [14]. In 2013, the first case report was published highlighting the success of LCAT replacement therapy using a recombinant enzyme form in a 53-year-old patient.

On the contrary, many gene therapy-based systems for the treatment of various genetic disorders have been developed and investigated during the last few years. Gene therapy promises the permanent expression of the functional protein after the incorporation of the corresponding gene into the host genome. However, as yet, gene therapy has not found wide clinical application, because, depending on the vectors used, it can be associated with risks for the patient, such as insertional mutagenesis, carcinogenic effects, immune responses, low gene-transfer efficiency, or protein misfolding [15].

3. mRNA as a novel therapeutic option

3.1. mRNA as a therapeutic agent

Next to gene therapy, wherein genetic defects are corrected by the introduction of specific DNA sequences into the genome, mRNA-based therapy promises new advances in the treatment of single-gene disorders. Via the delivery of a specific *in vitro*-generated mRNA to the target cells, the expression of a desired protein can be induced. The idea of using specific mRNAs to produce a protein of interest instead of protein replacement via DNA gene therapy was described 25 years ago by Wolff and colleagues [16]. At that time, however, the stability of the mRNA was poor and the immunogenicity was too high for the therapy to be practical. In recent years, the administration of the mRNA as a therapeutic agent has gained enormous potential in the fields of disease treatment, regenerative medicine, and vaccination [17–21].

Especially for monogenic diseases, mRNA therapy can be a highly beneficial alternative to classical gene therapy. Because monogenic diseases result in defective or missing protein synthesis, protein replacement therapy is primarily used for these kinds of diseases. The mRNA can be easily produced in large amounts for the protein of interest in comparison to pDNA, and if the mRNA is used, there is no need to integrate promoter and terminator regions in the sequence [22].

Moreover, the mRNA-based therapeutic strategy has significant advantages:

1. The introduced mRNA is translated by the cell's own translational machinery under physiological conditions.
2. By introducing a specific mRNA, the physiological state of the cell is not altered, because the effect is transient and not mutagenic.
3. The protein synthesis can be controlled directly without intervening in the human genome.
4. The mRNA does not need to enter the nucleus for translation [15, 20, 22].

Overall, this therapeutic strategy could be safely used in patients, and it is more cost-effective and easier to manipulate than gene therapy.

The standard *in vitro* procedure for mRNA generation begins with the plasmid, which contains the coding sequence of the protein. First, this sequence is amplified using polymerase chain reaction (PCR). The generated DNA template contains all the important elements of the mRNA. Second, the amplified PCR product is used to generate the mRNA. Therefore, *in vitro* transcription (IVT) is performed using the T7 or SP6 RNA polymerase, which synthesizes the mRNA. After the purification and quality control steps, the mRNA is ready to use (**Figure 1**).

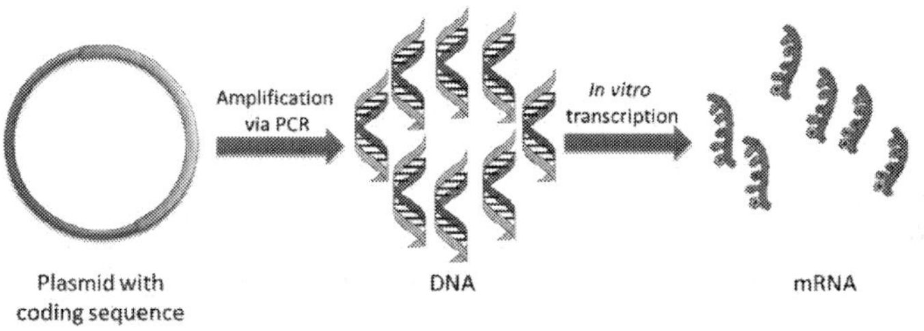

Figure 1. Generation process of the mRNA.

The mRNA is a single-stranded molecule containing a poly(A)-tail at the end and a 5′-cap at the beginning. The coding sequence for the protein is marked by a start codon and a stop codon. The untranslated regions (UTRs) are in between the cap/tail and the coding sequence (**Figure 2**) [16].

Figure 2. Schematic overview of the general structure of an IVT mRNA.

However, IVT mRNA is very sensitive to degradation by nucleases, which limits its suitability for transfections and therapeutic applications [21]. If the mRNA will be applied as a therapeutic agent, there is a need for special modifications. The modifications should be nonmutagenic and should not interfere with the translation machinery of the cell [23]. To improve the translation of the mRNA, the cap and the poly(A)-tail are important. The mRNA cap is responsible for the recognition of the ribosomes and binding to the ribosomes and for the initiation of the translation machinery [24]. It was reported that part of the unmodified cap structure could be bound in the wrong direction to the mRNA during IVT. Thus, the mRNA cannot be located by the cap-binding complex (CBC) of the ribosomes and the translation cannot follow [25]. By using an anti-reverse cap analog (ARCA) during IVT, this can be prevented. An ARCA contains a 5′-5′ triphosphate bridge, and the 3′OH is replaced by OCH_3, making it impossible for the cap to bind in the wrong direction [26]. The 5′-5′ bridge and the chemical modifications at the 3′-position lead to translatable mRNA of high quality and great translation efficiency [27, 28]. Furthermore, ARCA-capped mRNA is more resistant to hydrolases [29]. The poly(A)-tail, which binds to the polyadenosyl-binding-protein (PABP), is also very important for the stability and translation of the mRNA [30, 31]. The PABP interacts with the CBC and makes a circular structure of the mRNA molecule [15, 32]. This circular structure minimizes the contact surface for nucleases [33]. If the poly(A)-tail is shorter than 12 adenine nucleosides or removed completely, the cap structure is cleaved and the mRNA is degraded. Thus, the poly(A)-tail is very important to obtaining the cap structure and delaying the degradation of the mRNA [15, 34]. A poly(A)-tail of more than 60 adenine monophosphates increases the translation efficiency of the mRNA [35]. Therefore, the poly(A)-tail and the cap structure contribute to the stability of the mRNA [27, 36]. The 5′- and 3′-UTRs include specific regulatory sequence regions that are necessary to modulate the stability and translation of the mRNA [16]. The 3′-UTR region contains α- and β-globin sequences that enhance the stability and translation of the mRNA [16, 37, 38]. Furthermore, the 3′- and 5′-UTR regions inhibit the decapping and degradation of the mRNA [16, 39, 40]. On the contrary, if limited protein production is desired, faster mRNA degradation is possible by integrating AU-rich areas in the 3′-UTR region [41]. The protein-coding region of the mRNA can be optimized in two ways. First, the use of optimized codons leads to improved translation of the sequence to the desired protein. Second, via the optimization of the bases, the endonucleolytic degradation can be reduced [16].

3.2. Potential immune response against IVT mRNA

In comparison to recombinant protein in the substitution therapy approach, mRNA-translated proteins are autologously produced by the cell's own machinery. Usually, they undergo the correct modification and folding [16], but protein aggregation and immune reactions to the translated protein cannot be excluded. Therefore, it is necessary to investigate the activation of the immune system and distribution of the antibodies against IVT mRNA-translated proteins in clinical studies. Furthermore, the combination of the immune reaction to the translated mRNA protein and the foreign mRNA within the cells can lead to immunopathology in the cells or organs [16, 42].

Regarding protein replacement therapies, the activation of the immune system caused by mRNA administration might be disadvantageous. It is well known that foreign mRNAs as well as pDNAs activate the immune system via recognition through toll-like receptors (TLRs). TLRs recognize different pathogen patterns resulting in the expression of different cytokines. TLR3 and TLR7/8 are responsible for the recognition of different RNA types [43]. The mRNA and other RNA types, such as small interfering RNA (siRNA) or double-stranded RNA (dsRNA), are recognized by TLR3 [44–46], whereas TLR7/8 is activated by single-stranded RNA (ssRNA) [47]. In nonimmune cells, the recognition of the mRNA occurs through the retinoic acid-inducible gene I (RIG-I), which is activated by short RNA and dsRNA [48] and leads to interleukin-1β (IL-1β) production [49]. It is known that the IVT mRNA leads to a strong distribution of tumor necrosis factor-α (TNF-α) if the mRNA has no modifications [50]. Additionally, a strong type I interferon (type I IFN) response of the cells is induced upon contact with exogenous mRNAs. This is also induced by mRNAs that form secondary structures, such as loops or hairpins, or by mRNAs that bind to incompletely synthesized mRNA fragments or incompletely degraded DNA fragments [43, 51]. IFNs then activate the antiviral genes in the genome and lead to a translation stop and the degradation of the mRNA [43]. However, the immune reaction can be avoided if the mRNA is purified and if modified nucleotides are inserted during IVT [50, 52–54].

The incorporation of different modified nucleotides, such as pseudouridine (pseudo-U), 2′-thiouridine (2′-tU), 5′-methyluridin (5′-mU), 5-methylcytidine (5′-mC), or N^6-methyladenosine (N6-mA), can prevent the cellular immune response [50, 53]. These modified nucleotides incorporated in the mRNA help to avoid the activation of TLRs [50]. In particular, pseudo-U and 2′-tU make the recognition of IVT mRNA by RIG-I impossible [16, 54, 55]. To minimize the immune activation and optimize the translation efficiency of the mRNA to the protein, high-performance liquid chromatography (HPLC) purification should be applied to eliminate the dsRNA contamination that can still be present after the IVT process [56].

However, mRNA modification may also represent another risk. The naturally existing mRNA is degradable by the RNases in the extracellular space, but the modification of the IVT mRNA makes degradation more difficult [57]. Some of the modified nucleotides are associated with mitochondrial toxicities and hepatic failure and play a role in viral and tumor cell replication [58, 59]. Here too, further investigations and clinical trials are necessary to prove the risks and benefits of IVT mRNA modifications.

3.3. mRNA delivery systems and specific targeting

For clinical mRNA applications, a specific delivery system is needed, because otherwise the delivery of the mRNA to the target cells is unguaranteed and inefficient [27]. Therefore, the development and engineering of safe and effective delivery vectors for mRNA therapy is inevitable [60]. Viral and nonviral vectors can be used to bring the mRNA into the cells. For the direct translation of the mRNA to the protein, only positive-stranded viruses can be applied [61]. Negative-stranded viruses are not infectious and need an RNA-dependent RNA polymerase for mRNA translation [62]. However, viral vectors have some limitations; for example, they can be carcinogenic [63], they can activate immune responses [64], they can be difficult to produce [65], and they have a limited packaging capacity [66].

Nonviral vectors have lower immunogenicity, they can deliver large genetic molecules, and they are easier to produce [60, 67, 68]. Nonviral delivery systems can be subdivided into direct and indirect delivery systems. Direct delivery is possible via electroporation or gene guns. Electroporation is an early and efficient method of transporting mRNA into cells [69], whereby electrical pulses make the cell membrane permeable for the entry of the mRNA into the cytosol. This method does not induce immune activation, which may occur when mRNA carriers are used [27, 69]. Transfection using a gene gun requires heavy metal particles to get the nucleic acid into the cell [70]. This method allows the delivery of the mRNA to mammalian organs with minimal damage and leads to transient protein expression in the target tissues [70]. Self-assembled complexation of negatively charged mRNAs and positively charged liposomes and polysomes to lipoplexes or polyplexes is the most widely used method of bringing mRNA into cells [60]. The lipoplexes can be taken up by the cells in two different ways. The first way is by the endocytosis of the lipoplexes, whereby 98% of the lipoplexes enter the cell. The second way is through the fusion of the cell membrane and the lipoplexes, which results in the uptake of the remaining mRNA [71]. After the release of the mRNA into the cytosol, the protein translation can begin. The encapsulation of the mRNA into liposomes is a rapid, transient, and cell cycle-independent delivery method [27]. The translation of the mRNA to the protein can be measured 1 h after the transfection in nondividing cells [72].

For *in vivo* applications, the perfect mRNA delivery vector has to overcome various barriers:

1. Protection against nucleases,
2. Avoidance of nonspecific interactions with proteins or cells,
3. Prevention of renal clearance,
4. Permission of extravasation to target the tissue of interest, and
5. Increased entry into the cell [60].

For systemic delivery, the lipid and polymer complexes show protective properties against nucleases, whereby the mRNA is protected and the stability is increased [73]. However, liposome complexes sometimes interact with serum proteins. Together, they form aggregates or clots and are cleared rapidly [74]. The conjugation of the complexes with polyethylene glycol (PEG) helps to inhibit the nonspecific uptake and attachment to serum proteins [75].

Nanoscale delivery systems (10–200 nm) enhance the uptake efficiency and reduce the systemic toxicity [75]. The use of PEG coating of the liposomes increases the blood circulation time and avoids the detection of liposomes by immune cells [75, 76]. The liposomes have many advantages, such as low batch-to-batch variability, easy synthesis, biocompatibility, and scalability, over many other delivery systems [75]. The liposome surface can be functionalized by conjugation to chemically reactive lipid head groups [77]. This property makes it possible to functionalize the surface with ligands and thus enhance the target delivery [75, 78].

Different methods of application are tested for *in vivo* delivery. Polyplex nanomicelles applied with hydrodynamic intravenous injection have been shown to effectively deliver the mRNA to the liver in mouse models. This method shows a strong protein expression in nearly 100% of liver cells [79]. Intramuscular or intraperitoneal injection of erythropoietin (EPO)-coding mRNA complexed with cationic lipids leads to significantly high levels of EPO *in vivo* [20, 80]. Intratracheal and intranasal applications of Foxp3 mRNA show protective properties against asthma in mice [81]. Furthermore, it has been shown that intradermal [82] and intranodal [83] applications of the mRNA in animal models resulted in immunization against tumors. Many other methods of applying the mRNA have been described in publications in recent years, and this field is developing rapidly [19, 84]. Furthermore, patient-centered applications have improved, especially for mRNA therapy. Nebulization with the Pari-Boy® is the standard method used to apply the drugs to the lung. A study on the influence on mRNA transfection efficiency shows that the nebulization of complexed mRNA has no effects *in vitro* [85].

http://dx.doi.org/10.5772/62243

3.4. mRNA applications

Gene therapy allows the replacement of a defective gene through substitution and integration of the correct genetic code in the genome. The genome integration of this genetic code via viruses guarantees highly efficient gene replacement methods. However, undesirable effects, such as mutagenesis and innate immune response, may jeopardize the life or safety of the patient [27, 86]. Nonviral gene delivery is safer, but it is also associated with lower transfection efficiency due to insufficient nuclear transport. Furthermore, modifications of the pDNA, including adding a strong constructive promoter to improve transcription efficiency, may lead to unexpected alterations in the genome [27, 86, 87].

Protein-substitution therapy is associated with adverse reactions, such as headache, dizziness, and nausea [88], and high costs [89]. Recombinant proteins have been expressed in different microorganisms [90–92], plant cells [93], and human cells [94–96], but they are linked to disadvantages such as nonglycosylation or incorrect glycosylation of the product as well as product contaminations with endotoxins. Plasma-derived proteins, which are purified from human donor blood, are limited and prone to contamination.

The mRNA is an alternative to overcome the disadvantages of pDNA and direct protein substitution [15]. The mRNA does not need to enter the nucleus for transcription, because the mRNA is directly translated in the cytosol of the cells; thus, the insertion of exogenous DNA into the genome poses no risk. Furthermore, the mRNA uses the cell's own translation machinery and requires no strong promoter. Effective mRNA transfer can also take place in nondividing cells, and immunogenicity can be overcome by different modifications of the mRNA molecule [27]. Moreover, compared to protein-substitution therapy, the expression of receptors and intracellular molecules can be induced with specific mRNAs (**Figure 3**).

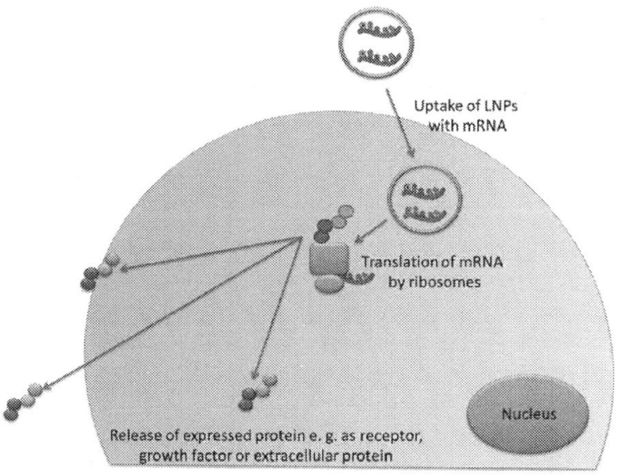

Figure 3. The way from encapsulated mRNA to protein expression in the target cell.

Several studies describe that the use of different mRNAs leads to an increase in the respective protein *in vitro* and *in vivo*. The first application of the mRNA was performed in 1992 [97]. However, the broad application of IVT mRNA only became possible as modified nucleotides were used; thus, a reduction of immune reactions and an increase of mRNA translation efficiency were achieved [50, 56]. For example, the successful expression of surfactant protein B (SP-B) of therapeutically relevant levels was shown in a mouse model after the application of SP-B-encoding mRNA via direct administration to the lung. The results showed that the inflammation of the lung (and the inevitable respiratory failure and death) was prevented [20, 98]. Likewise, the expression of the regulatory T-cell transcription factor (FOXP3) led to the prevention of asthma in a mouse model. After the administration of FOXP3-coding mRNA to the lung, the expressed protein protected the lung from allergen-induced inflammation. This mRNA approach can be used as a preventive and therapeutic drug [81]. In a different study, the expression of the AAT protein was shown after the transfection of cells with AAT-encoding mRNA. The level of AAT was measurable in the cellular supernatant 48 h after transfection, and the functionality of the protein was proven. All these studies show that mRNA therapy also promises a novel therapeutic strategy in the treatment of single-gene diseases such as AATD [99].

In the case of FH, it is also possible to induce the expression of functional LDLR after the transfection and thus regulate LDL metabolism. In this case, functional LDLR expression can help to prevent secondary diseases, such as stroke and atherosclerotic plaques. Although the induction of the expression of proteins, such as LCAT and FIX, can be performed similarly to the approach of AATD-mRNA therapy, different organs are targeted. Regarding hemophilia B and LCATD, hepatocytes should be transfected, which can be achieved by intravenous injections of the complexed protein-encoding mRNA.

Depending on the application method and the genetic defect, it is important that the cell type of interest is targeted and transfected with IVT mRNA. Additionally, the dose-effect relationship could be a challenge *in vivo*, whereby the bioavailability and individual variations also play a pivotal role and potentially make individual dose adaptation necessary [16]. Overall, mRNA application promises an effective and low-cost therapeutic strategy with the potential to efficiently correct serious monogenic diseases.

3.5. Summary

Overall, the mRNA as a novel therapy for monogenic diseases has many advantages over the currently existing treatment options such as substitution therapy or gene therapy. Using the mRNA, the expression of nearly all proteins can be induced. In this way, not only extracellular proteins, but also proteins, which are important for the function of the cells, such as growth factors or receptors can be generated. Furthermore, the mRNA does not need to penetrate into the nucleus. The effect of the mRNA is transient, thereby enabling the precise control of protein expression. Depending on the introduced modifications, the half-life of the respective mRNA and thus a reduction or increase in mRNA stability can be determined. Developments in recent years, such as modification and purification methods, have made it possible to use the mRNA as a therapeutic agent, because it became possible to control the immune reaction, which is typically triggered by exogenous mRNA. Further developments increased the extracellular stability and made mRNA transfection of the cells with nonviral vectors possible and efficient. The mRNA can be produced in large batches and under good manufacturing practice (GMP) conditions without batch-to-batch variations. Thereby, the production of specific mRNA is significantly cheaper compared to the production of the corresponding protein for substitution therapy.

The mRNA as a therapeutic agent could be a great help for patients suffering from monogenic diseases. The flexibility and variability of proteins that can be replaced by the cell's own translational machinery through the use of the mRNA is nearly unlimited. This makes mRNA to a unique therapeutic molecule, which will revolutionize therapeutic options for affected patients in the coming years.

Author details

Tatjana Michel, Hans-Peter Wendel and Stefanie Krajewski*

*Address all correspondence to: stefanie.krajewski@uni-tuebingen.de

Department of Thoracic, Cardiac and Vascular Surgery, University of Tuebingen, Tuebingen, Germany

References

[1] Collins FS, Lander ES, Rogers J, Waterston RH, Conso IHGS: Finishing the euchromatic sequence of the human genome. *Nature* 2004, 431(7011):931–945.

[2] Antonarakis SE, Beckmann JS: Mendelian disorders deserve more attention. *Nat Rev Genet* 2006, 7(4):277–282.

[3] Ropers HH: Single gene disorders come into focus—again. *Dialogues Clin Neurosci* 2010, 12(1):95–102.

[4] Laurell CB, Eriksson S: The electrophoretic α1-globulin pattern of serum in α1-antitrypsin deficiency. 1963. *COPD* 2013, 10 Suppl 1:3–8.

[5] Hurley K, Lacey N, O'Dwyer CA, Bergin DA, McElvaney OJ, O'Brien ME, McElvaney OF, Reeves EP, McElvaneyNG: α-1 antitrypsin augmentation therapy corrects accelerated neutrophil apoptosis in deficient individuals. *J Immunol* 2014, 193(8):3978–3991.

[6] Cancio MI, Reiss UM, Nathwani AC, Davidoff AM, Gray JT: Developments in the treatment of hemophilia B: focus on emerging gene therapy. *Appl Clin Genet* 2013, 6:91–101.

[7] Francone OL, Gurakar A, Fielding C: Distribution and functions of lecithin: cholesterol acyltransferase and cholesteryl ester transfer protein in plasma lipoproteins. Evidence for a functional unit containing these activities together with apolipoproteins A-I and D that catalyzes the esterification and transfer of cell-derived cholesterol. *J Biol Chem* 1989, 264(12):7066–7072.

[8] Kuivenhoven JA, Pritchard H, Hill J, Frohlich J, Assmann G, Kastelein J: The molecular pathology of lecithin:cholesterol acyltransferase (LCAT) deficiency syndromes. *J Lipid Res* 1997, 38(2):191–205.

[9] McIntyre N: Familial LCAT deficiency and fish-eye disease. *J Inherited Metab Dis* 1988, 11(Suppl 1):45–56.

[10] Raal FJ, Santos RD: Homozygous familial hypercholesterolemia: current perspectives on diagnosis and treatment. *Atherosclerosis* 2012, 223(2):262–268.

[11] Pijlman AH, Huijgen R, Verhagen SN, Imholz BP, Liem AH, Kastelein JJ, Abbink EJ, Stalenhoef AF, Visseren FL: Evaluation of cholesterol lowering treatment of patients with familial hypercholesterolemia: a large cross-sectional study in The Netherlands. *Atherosclerosis* 2010, 209(1):189–194.

[12] American Thoracic Society, European Respiratory Society: American Thoracic Society/European Respiratory Society statement: standards for the diagnosis and management of individuals with α-1 antitrypsin deficiency. *Am J Respir Crit Care Med* 2003, 168(7): 818–900.

[13] Petrache I, Hajjar J, Campos M: Safety and efficacy of α-1-antitrypsin augmentation therapy in the treatment of patients with α-1-antitrypsin deficiency. *Biol Targets Ther* 2009, 3:193–204.

[14] Kunnen S, Van Eck M: Lecithin:cholesterol acyltransferase: old friend or foe in atherosclerosis? *J Lipid Res* 2012, 53(9):1783–1799.

[15] Yamamoto A, Kormann M, Rosenecker J, Rudolph C: Current prospects for mRNA gene delivery. *Eur J Pharm Biopharm* 2009, 71(3):484–489.

[16] Sahin U, Kariko K, Tureci O: mRNA-based therapeutics—developing a new class of drugs. *Nat Rev Drug Discov* 2014, 13(10):759–780.

[17] Kübler H, Maurer T, Stenzl A, Feyerabend S, Steiner U, Schostak M, Schultze-Seemann W, vom Dorp F, Pilla L, Viatali G, et al.: Final analysis of a phase I/IIa study with CV9103, an intradermally administered prostate cancer immunotherapy based on self-adjuvanted mRNA. *J Clin Oncol* 2011, 29:498–499.

[18] Fotin-Mleczek M, Duchardt KM, Lorenz C, Pfeiffer R, Ojkic-Zrna S, Probst J, Kallen KJ: Messenger RNA-based vaccines with dual activity induce balanced TLR-7 dependent adaptive immune responses and provide antitumor activity. *J Immunother* 2011, 34(1): 1–15.

[19] Petsch B, Schnee M, Vogel AB, Lange E, Hoffmann B, Voss D, Schlake T, Thess A, Kallen KJ, Stitz L, et al.: Protective efficacy of *in vitro* synthesized, specific mRNA vaccines against influenza A virus infection. *Nat Biotechnol* 2012, 30(12):1210–1216.

[20] Kormann MS, Hasenpusch G, Aneja MK, Nica G, Flemmer AW, Herber-Jonat S, Huppmann M, Mays LE, Illenyi M, Schams A, et al.: Expression of therapeutic proteins after delivery of chemically modified mRNA in mice. *Nat Biotechnol* 2011, 29(2):154–157.

[21] Avci-Adali M, Behring A, Keller T, Krajewski S, Schlensak C, Wendel HP: Optimized conditions for successful transfection of human endothelial cells with *in vitro* synthesized and modified mRNA for induction of protein expression. *J Biol Eng* 2014, 8(1):8.

[22] Tavernier G, Andries O, Demeester J, Sanders NN, De Smedt SC, Rejman J: mRNA as gene therapeutic: how to control protein expression. *J Control Release* 2011, 150(3):238–247.

[23] Wang X, He C: Dynamic RNA modifications in posttranscriptional regulation. *Mol Cell* 2014, 56(1):5–12.

[24] Gallie DR: The cap and poly(A) tail function synergistically to regulate mRNA translational efficiency. *Genes Dev* 1991, 5(11):2108–2116.

[25] Matsuo H, Li H, McGuire AM, Fletcher CM, Gingras AC, Sonenberg N, Wagner G: Structure of translation factor eIF4E bound to m7GDP and interaction with 4E-binding protein. *Nat Struct Biol* 1997, 4(9):717–724.

[26] Stepinski J, Waddell C, Stolarski R, Darzynkiewicz E, Rhoads RE: Synthesis and properties of mRNAs containing the novel "anti-reverse" cap analogs 7-methyl(3′-O-methyl)GpppG and 7-methyl (3′-deoxy)GpppG. *RNA* 2001, 7(10):1486–1495.

[27] Youn H, Chung JK: Modified mRNA as an alternative to plasmid DNA (pDNA) for transcript replacement and vaccination therapy. *Expert Opin Biol Ther* 2015, 15(9):1337–1348.

[28] Mockey M, Goncalves C, Dupuy FP, Lemoine FM, Pichon C, Midoux P: mRNA transfection of dendritic cells: synergistic effect of ARCA mRNA capping with poly(A) chains in cis and in trans for a high protein expression level. *Biochem Biophys Res Commun* 2006, 340(4):1062–1068.

[29] Grudzien E, Kalek M, Jemielity J, Darzynkiewicz E, Rhoads RE: Differential inhibition of mRNA degradation pathways by novel cap analogs. *J Biol Chem* 2006, 281(4):1857–1867.

[30] Chang H, Lim J, Ha M, Kim VN: TAIL-seq: genome-wide determination of poly(A) tail length and 3′ end modifications. *Mol Cell* 2014, 53(6):1044–1052.

[31] Peng J, Murray EL, Schoenberg DR: *In vivo* and *in vitro* analysis of poly(A) length effects on mRNA translation. *Methods Mol Biol* 2008, 419:215–230.

[32] Gallie DR: A tale of two termini: a functional interaction between the termini of an mRNA is a prerequisite for efficient translation initiation. *Gene* 1998, 216(1):1–11.

[33] Newbury SF: Control of mRNA stability in eukaryotes. *Biochem Soc Trans* 2006, 34(Pt 1):30–34.

[34] Schwartz D, Decker CJ, Parker R: The enhancer of decapping proteins, Edc1p and Edc2p, bind RNA and stimulate the activity of the decapping enzyme. *RNA* 2003, 9(2): 239–251.

[35] Elango N, Elango S, Shivshankar P, Katz MS: Optimized transfection of mRNA transcribed from a d(A/T)100 tail-containing vector. *Biochem Biophys Res Commun* 2005, 330(3):958–966.

[36] Parker R, ShethU: P bodies and the control of mRNA translation and degradation. *Mol Cell* 2007, 25(5):635–646.

[37] Kariko K, Kuo A, Barnathan E: Overexpression of urokinase receptor in mammalian cells following administration of the *in vitro* transcribed encoding mRNA. *Gene Ther* 1999, 6(6):1092–1100.

[38] Ross J, Sullivan TD: Half-lives of β and γ globin messenger RNAs and of protein synthetic capacity in cultured human reticulocytes. *Blood* 1985, 66(5):1149–1154.

[39] Kuhn AN, Beibetaert T, Simon P, Vallazza B, Buck J, Davies BP, Tureci O, Sahin U: mRNA as a versatile tool for exogenous protein expression. *Curr Gene Ther* 2012, 12(5): 347–361.

[40] Bergman N, Moraes KC, Anderson JR, Zaric B, Kambach C, Schneider RJ, Wilusz CJ, Wilusz J: Lsm proteins bind and stabilize RNAs containing 5′ poly(A) tracts. *Nat Struct Mol Biol* 2007, 14(9):824–831.

[41] Chen CY, Shyu AB: AU-rich elements: characterization and importance in mRNA degradation. *Trends Biochem Sci* 1995, 20(11):465–470.

[42] Racanelli V, Rehermann B: The liver as an immunological organ. *Hepatology* 2006, 43(2 Suppl 1):S54–S62.

[43] Janeway C: Immunologie, 7. Aufl. edn. Heidelberg/Berlin: Spektrum, Akadem. Verl.; 2009.

[44] Kariko K, Ni H, Capodici J, Lamphier M, Weissman D: mRNA is an endogenous ligand for Toll-like receptor 3. *J Biol Chem* 2004, 279(13):12542–12550.

[45] Alexopoulou L, Holt AC, Medzhitov R, Flavell RA: Recognition of double-stranded RNA and activation of NF-κB by Toll-like receptor 3. *Nature* 2001, 413(6857):732–738.

[46] Kariko K, Bhuyan P, Capodici J, Weissman D: Small interfering RNAs mediate sequence-independent gene suppression and induce immune activation by signaling through toll-like receptor 3. *J Immunol* 2004, 172(11):6545–6549.

[47] Heil F, Hemmi H, Hochrein H, Ampenberger F, Kirschning C, Akira S, Lipford G, Wagner H, Bauer S: Species-specific recognition of single-stranded RNA via toll-like receptor 7 and 8. *Science* 2004, 303(5663):1526–1529.

[48] Schlee M, Roth A, Hornung V, Hagmann CA, Wimmenauer V, Barchet W, Coch C, Janke M, Mihailovic A, Wardle G, et al.: Recognition of 5′ triphosphate by RIG-I helicase requires short blunt double-stranded RNA as contained in panhandle of negative-strand virus. *Immunity* 2009, 31(1):25–34.

[49] Poeck H, Bscheider M, Gross O, Finger K, Roth S, Rebsamen M, Hannesschlager N, Schlee M, Rothenfusser S, Barchet W, et al.: Recognition of RNA virus by RIG-I results in activation of CARD9 and inflammasome signaling for interleukin 1β production. *Nat Immunol* 2010, 11(1):63–69.

[50] Kariko K, Buckstein M, Ni H, Weissman D: Suppression of RNA recognition by Toll-like receptors: the impact of nucleoside modification and the evolutionary origin of RNA. *Immunity* 2005, 23(2):165–175.

[51] Rautsi O, Lehmusvaara S, Salonen T, Hakkinen K, Sillanpaa M, Hakkarainen T, Heikkinen S, Vahakangas E, Yla-Herttuala S, Hinkkanen A, et al.: Type I interferon response against viral and non-viral gene transfer in human tumor and primary cell lines. *J Gene Med* 2007, 9(2):122–135.

[52] Kariko K, Weissman D: Naturally occurring nucleoside modifications suppress the immunostimulatory activity of RNA: implication for therapeutic RNA development. *Curr Opin Drug Discov Dev* 2007, 10(5):523–532.

[53] Kariko K, Muramatsu H, Welsh FA, Ludwig J, Kato H, Akira S, Weissman D: Incorporation of pseudouridine into mRNA yields superior nonimmunogenic vector with increased translational capacity and biological stability. *Mol Ther* 2008, 16(11):1833–1840.

[54] Anderson BR, Muramatsu H, Nallagatla SR, Bevilacqua PC, Sansing LH, Weissman D, Kariko K: Incorporation of pseudouridine into mRNA enhances translation by diminishing PKR activation. *Nucleic Acids Res* 2010, 38(17):5884–5892.

[55] Hornung V, Ellegast J, Kim S, Brzozka K, Jung A, Kato H, Poeck H, Akira S, Conzelmann KK, Schlee M, et al.: 5′-Triphosphate RNA is the ligand for RIG-I. *Science* 2006, 314(5801):994–997.

[56] Kariko K, Muramatsu H, Ludwig J, Weissman D: Generating the optimal mRNA for therapy: HPLC purification eliminates immune activation and improves translation of nucleoside-modified, protein-encoding mRNA. *Nucleic Acids Res* 2011, 39(21):e142.

[57] Dyer KD, Rosenberg HF: The RNase a superfamily: generation of diversity and innate host defense. *Mol Diversity* 2006, 10(4):585–597.

[58] McKenzie R, Fried MW, Sallie R, Conjeevaram H, Di Bisceglie AM, Park Y, Savarese B, Kleiner D, Tsokos M, Luciano C, et al.: Hepatic failure and lactic acidosis due to

fialuridine (FIAU), an investigational nucleoside analogue for chronic hepatitis B. *N Engl J Med* 1995, 333(17):1099–1105.

[59] Lewis W: Defective mitochondrial DNA replication and NRTIs: pathophysiological implications in AIDS cardiomyopathy. *Am J Physiol Heart Circ Physiol* 2003, 284(1):H1–H9.

[60] Yin H, Kanasty RL, Eltoukhy AA, Vegas AJ, Dorkin JR, Anderson DG: Non-viral vectors for gene-based therapy. *Nat Rev Genet* 2014, 15(8):541–555.

[61] Agapov EV, Frolov I, Lindenbach BD, Pragai BM, Schlesinger S, Rice CM: Noncytopathic Sindbis virus RNA vectors for heterologous gene expression. *Proc Natl Acad Sci U S A* 1998, 95(22):12989–12994.

[62] Bitzer M, Armeanu S, Lauer UM, Neubert WJ: Sendai virus vectors as an emerging negative-strand RNA viral vector system. *J Gene Med* 2003, 5(7):543–553.

[63] Baum C, Kustikova O, Modlich U, Li Z, Fehse B: Mutagenesis and oncogenesis by chromosomal insertion of gene transfer vectors. *Hum Gene Ther* 2006, 17(3):253–263.

[64] Bessis N, GarciaCozar FJ, Boissier MC: Immune responses to gene therapy vectors: influence on vector function and effector mechanisms. *Gene Ther* 2004, 11 Suppl 1:S10–S17.

[65] Bouard D, Alazard-Dany D, Cosset FL: Viral vectors: from virology to transgene expression. *Br J Pharmacol* 2009, 157(2):153–165.

[66] Thomas CE, Ehrhardt A, Kay MA: Progress and problems with the use of viral vectors for gene therapy. *Nat Rev Genet* 2003, 4(5):346–358.

[67] Pack DW, Hoffman AS, Pun S, Stayton PS: Design and development of polymers for gene delivery. *Nat Rev Drug Discov* 2005, 4(7):581–593.

[68] Mintzer MA, Simanek EE: Nonviral vectors for gene delivery. *Chem Rev* 2009, 109(2): 259–302.

[69] Van Tendeloo VF, Ponsaerts P, Lardon F, Nijs G, Lenjou M, Van Broeckhoven C, Van Bockstaele DR, Berneman ZN: Highly efficient gene delivery by mRNA electroporation in human hematopoietic cells: superiority to lipofection and passive pulsing of mRNA and to electroporation of plasmid cDNA for tumor antigen loading of dendritic cells. *Blood* 2001, 98(1):49–56.

[70] Sohn RL, Murray MT, Schwarz K, Nyitray J, Purray P, Franko AP, Palmer KC, Diebel LN, Dulchavsky SA: *In-vivo* particle mediated delivery of mRNA to mammalian tissues: ballistic and biologic effects. *Wound Repair Regen* 2001, 9(4):287–296.

[71] Zhdanov RI, Podobed OV, Vlassov VV: Cationic lipid-DNA complexes-lipoplexes-for gene transfer and therapy. *Bioelectrochemistry* 2002, 58(1):53–64.

[72] Zou S, Scarfo K, Nantz MH, Hecker JG: Lipid-mediated delivery of RNA is more efficient than delivery of DNA in non-dividing cells. *Int J Pharm* 2010, 389(1–2):232–243.

[73] Luo D, Saltzman WM: Synthetic DNA delivery systems. *Nat Biotechnol* 2000, 18(1):33–37.

[74] Tros de Ilarduya C, Duzgunes N: Efficient gene transfer by transferrin lipoplexes in the presence of serum. *Biochim Biophys Acta* 2000, 1463(2):333–342.

[75] Noble GT, Stefanick JF, Ashley JD, Kiziltepe T, Bilgicer B: Ligand-targeted liposome design: challenges and fundamental considerations. *Trends Biotechnol* 2014, 32(1):32–45.

[76] Allen TM, Cullis PR: Liposomal drug delivery systems: from concept to clinical applications. *Adv Drug Deliv Rev* 2013, 65(1):36–48.

[77] Andresen TL, Jensen SS, Jorgensen K: Advanced strategies in liposomal cancer therapy: problems and prospects of active and tumor specific drug release. *Prog Lipid Res* 2005, 44(1):68–97.

[78] Ruoslahti E: Peptides as targeting elements and tissue penetration devices for nanoparticles. *Adv Mater* 2012, 24(28):3747–3756.

[79] Matsui A, Uchida S, Ishii T, Itaka K, Kataoka K: Messenger RNA-based therapeutics for the treatment of apoptosis-associated diseases. *Sci Rep* 2015, 5:15810.

[80] Kariko K, Muramatsu H, Keller JM, Weissman D: Increased erythropoiesis in mice injected with submicrogram quantities of pseudouridine-containing mRNA encoding erythropoietin. *Mol Ther* 2012, 20(5):948–953.

[81] Mays LE, Ammon-Treiber S, Mothes B, Alkhaled M, Rottenberger J, Muller-Hermelink ES, Grimm M, Mezger M, Beer-Hammer S, von Stebut E, et al.: Modified Foxp3 mRNA protects against asthma through an IL-10-dependent mechanism. *J Clin Invest* 2013, 123(3):1216–1228.

[82] Granstein RD, Ding W, Ozawa H: Induction of anti-tumor immunity with epidermal cells pulsed with tumor-derived RNA or intradermal administration of RNA. *J Invest Dermatol* 2000, 114(4):632–636.

[83] Kreiter S, Selmi A, Diken M, Koslowski M, Britten CM, Huber C, Tureci O, Sahin U: Intranodal vaccination with naked antigen-encoding RNA elicits potent prophylactic and therapeutic antitumoral immunity. *Cancer Res* 2010, 70(22):9031–9040.

[84] Azarmi S, Roa WH, Lobenberg R: Targeted delivery of nanoparticles for the treatment of lung diseases. *Adv Drug Deliv Rev* 2008, 60(8):863–875.

[85] Johler SM, Rejman J, Guan S, Rosenecker J: Nebulisation of IVT mRNA complexes for intrapulmonary administration. *PLoS One* 2015, 10(9):e0137504.

[86] Sheridan C: Gene therapy finds its niche. *Nat Biotechnol* 2011, 29(2):121–128.

[87] Das SK, Menezes ME, Bhatia S, Wang XY, Emdad L, Sarkar D, Fisher PB: Gene therapies for cancer: strategies, challenges and successes. *J Cell Physiol* 2015, 230(2):259–271.

[88] Stoller JK, Fallat R, Schluchter MD, O'Brien RG, Connor JT, Gross N, O'Neil K, Sandhaus R, CrystalRG: Augmentation therapy with α1-antitrypsin: patterns of use and adverse events. *Chest* 2009, 136(5 Suppl):e30.

[89] Silverman EK, Sandhaus RA: Clinical practice. α1-Antitrypsin deficiency. *N Engl J Med* 2009, 360(26):2749–2757.

[90] Chill L, Trinh L, Azadi P, Ishihara M, Sonon R, Karnaukhova E, Ophir Y, Golding B, Shiloach J: Production, purification, and characterization of human α1 proteinase inhibitor from *Aspergillus niger*. *Biotechnol Bioeng* 2009, 102(3):828–844.

[91] Courtney M, Buchwalder A, Tessier LH, Jaye M, Benavente A, Balland A, Kohli V, Lathe R, Tolstoshev P, Lecocq JP: High-level production of biologically active human α1-antitrypsin in Escherichia coli. *Proc Natl Acad Sci U S A* 1984, 81(3):669–673.

[92] Kwon KS, Song M, Yu MH: Purification and characterization of α1-antitrypsin secreted by recombinant yeast *Saccharomyces diastaticus*. *J Biotechnol* 1995, 42(3):191–195.

[93] Huang J, Sutliff TD, Wu L, Nandi S, Benge K, Terashima M, Ralston AH, Drohan W, Huang N, Rodriguez RL: Expression and purification of functional human α1-antitrypsin from cultured plant cells. *Biotechnol Prog* 2001, 17(1):126–133.

[94] Ross D, Brown T, Harper R, Pamarthi M, Nixon J, Bromirski J, Li CM, Ghali R, Xie H, Medvedeff G, et al.: Production and characterization of a novel human recombinant α-1-antitrypsin in PER.C6 cells. *J Biotechnol* 2012, 162(2–3):262–273.

[95] Niklas J, Priesnitz C, Rose T, Sandig V, Heinzle E: Metabolism and metabolic burden by α1-antitrypsin production in human AGE1.HN cells. *Metab Eng* 2013, 16:103–114.

[96] Blanchard V, Liu X, Eigel S, Kaup M, Rieck S, Janciauskiene S, Sandig V, Marx U, Walden P, Tauber R, et al.: N-glycosylation and biological activity of recombinant human α1-antitrypsin expressed in a novel human neuronal cell line. *Biotechnol Bioeng* 2011, 108(9):2118–2128.

[97] Jirikowski GF, Sanna PP, Maciejewski-Lenoir D, Bloom FE: Reversal of diabetes insipidus in Brattleboro rats: intrahypothalamic injection of vasopressin mRNA. *Science* 1992, 255(5047):996–998.

[98] Mahiny AJ, Dewerth A, Mays LE, Alkhaled M, Mothes B, Malaeksefat E, Loretz B, Rottenberger J, Brosch DM, Reautschnig P, et al.: *In vivo* genome editing using nuclease-encoding mRNA corrects SP-B deficiency. *Nat Biotechnol* 2015, 33(6):584–586.

[99] Michel T, Kankura A, Salinas Medina ML, Kurz J, Behring A, Avci-Adali M, Nolte A, Schlensak C, Wendel HP, Krajewski S: *In vitro* evaluation of a novel mRNA-based therapeutic strategy for the treatment of patients suffering from α-1-antitrypsin deficiency. *Nucleic Acids Ther* 2015, 25(5):235–244.

Chapter 2

Gene Delivery Technologies for Efficient Genome Editing: Applications in Gene Therapy

Francisco Martin, Sabina Sánchez-Hernández,
Alejandra Gutierrez-Guerrero and
Karim Benabdellah

Additional information is available at the end of the chapter

http://dx.doi.org/10.5772/62776

Abstract

Specific nucleases (SNs), including ZFNs, TALENs, and CRISPR (clustered regularly interspaced palindromic repeats), are powerful tools for genome editing (GE). These tools have achieved efficient gene repair and gene disruption of human primary cells. However, their efficiency and safety must be improved before translation into clinic. In particular, one of the main hurdles of GE technology is the delivery of the different components into the nucleus of target cells. Successful gene editing must be able to deliver the SNs and/or the donor DNA into a large number of target cells in order to have a therapeutic benefit. In addition, the delivery must be nontoxic and the SNs must be innocuous to the target cells. In this chapter, we will summarize the different ways to deliver SNs and donor DNA.

Keywords: gene edition, gene therapy, delivery, viral vectors, specific nucleases

1. Introduction

Although genome editing (GE) technologies have been used for more than 30 years, the efficiency and specificity was too low to be used in gene therapy (GT). However, the development of specific nucleases (SNs) that can enhance homologous directed recombination (HDR) up to 10,000 times and allows the generation of specific mutations have open new possibilities for the use of GE for GT applications. SNs can create specific DSBs at target locations in the genome. These DSBs must be repaired by the cell's endogenous mechanisms by either HDR, a high-fidelity reparation process, or by nonhomologous end joining (NHEJ), an error prone process that results in insertions or deletions (indels) at the cleavage site. This repair mechanism is used as a platform for GE. Depending on the mechanism used by the cells to repair the DSBs, we can repair, insert, or delete DNA fragments in the genome of the target cells. There are four families of SNs being used: meganucleases (MNs) [1], zinc finger nucleases (ZFNs) [2], transcription activator-like effectors nucleases (TALENs) [3], and the clustered regularly interspaced short palindromic repeats (CRISPR) and CRISPR Associated 9 (Cas9) (CRISPR/Cas9) system [4] also named RNA-guided nulceases (RGNs). MNs, ZFNs, and TALENs used the principles of DNA-protein recognition to target specific locus. However, the difficulties of protein design, synthesis, and validation blocked the widespread adoption of these engineered nucleases for routine use in many laboratories. The field is experiencing a new phase thanks to the development of the RGNs [4]. This two-component system can achieve specific cleavage in any target DNA location guided by a small RNA molecule named gRNA [5].

However, in spite of the great advances in SNs design, the translation of GE technologies into clinic still requires several refinements both in terms of specificity and efficiency [6]. The efficiency of a particular GE strategy largely relies on the efficiency of delivery of the SNs and/or the donor DNA. In this chapter, we will discuss the different tools available for the delivery of the different components required for GE.

2. Viral-based vectors

GE using SNs and/or a donor DNA requires vector systems that efficiently deliver both components into the target cells. This is a relative easy task for cells that are growing in the laboratories but is a much difficult task when we target primary human cells. In general, viral-based vectors are more efficient than non-viral for most primary human cells. Since transient expression of GE components is preferred over stable expression, we will focus on episomal (non-integrative) viral-based vectors.

2.1. Adenoassociated virus (AAV)-based vectors

AAVs have been used for the delivery of ZFNs; however, the limited capacity of the AAV vectors makes difficult the use of these vectors for the delivery of Cas9 and TALEN. The first report showing the efficacy of AAV vectors to deliver ZFNs was published in 2012 [7, 8]. Soon later, the group of Katherine High showed that systemic AAV-ZFNs and AAV-corrective donor template enables production of high levels of human factor IX in a murine model of hemophilia B [9]. In another study, Weber et al. [10] reported that the administration of AAV-ZFNs-targeting HBV polymerase achieved an inhibition of HBV replication. In spite of their big size, several groups have attempted to deliver *S. pyogenes* Cas9 and its gRNA using AAV vectors [11]. However, oversized AAV vectors render inconsistent results [12]. As an alternative, different groups deliver the gRNA and the Cas9 in separate AAVs with very promising

[23] Sharma, R., et al., *In vivo genome editing of the albumin locus as a platform for protein replacement therapy*. Blood, 2015. 126(15): pp. 1777–1784.

[24] Maggio, I., and Goncalves, M.A., *Genome editing at the crossroads of delivery, specificity, and fidelity*. Trends Biotechnol, 2015. 33(5): pp. 280–291.

[25] Ohbayashi, F., et al., *Correction of chromosomal mutation and random integration in embryonic stem cells with helper-dependent adenoviral vectors*. Proc Natl Acad Sci U S A, 2005. 102(38): pp. 13628–13633.

[26] Liu, G.H., et al., *Targeted gene correction of laminopathy-associated LMNA mutations in patient-specific iPSCs*. Cell Stem Cell, 2010. 8(6): pp. 688–694.

[27] Perez, E.E., et al., *Establishment of HIV-1 resistance in CD4+ T cells by genome editing using zinc-finger nucleases*. Nat Biotechnol, 2008. 26(7): pp. 808–816.

[28] Maier, D.A., et al., *Efficient clinical scale gene modification via zinc finger nuclease-targeted disruption of the HIV co-receptor CCR5*. Hum Gene Ther, 2013. 24(3): pp. 245–258.

[29] Tebas, P., et al., *Gene editing of CCR5 in autologous CD4 T cells of persons infected with HIV*. N Engl J Med, 2014. 370(10): pp. 901–910.

[30] Holkers, M., Cathomen, T., and Goncalves, M.A., *Construction and characterization of adenoviral vectors for the delivery of TALENs into human cells*. Methods, 2014. 69(2): pp. 179–187.

[31] Cheng, R., et al., *Efficient gene editing in adult mouse livers via adenoviral delivery of CRISPR/Cas9*. FEBS Lett, 2014. 588(21): pp. 3954–3958.

[32] Li, C., et al., *Inhibition of HIV-1 infection of primary CD4+ T-cells by gene editing of CCR5 using adenovirus-delivered CRISPR/Cas9*. J Gen Virol, 2015. 96(8): pp. 2381–2393.

[33] Holkers, M., et al., *Adenoviral vector DNA for accurate genome editing with engineered nucleases*. Nat Methods, 2014. 11(10): pp. 1051–1057.

[34] Yanez-Munoz, R.J., et al., *Effective gene therapy with nonintegrating lentiviral vectors*. Nat Med, 2006. 12(3): pp. 348–353.

[35] Lombardo, A., et al., *Gene editing in human stem cells using zinc finger nucleases and integrase-defective lentiviral vector delivery*. Nat Biotechnol, 2007. 25(11): pp. 1298–1306.

[36] Cornu, T.I., and Cathomen, T., *Targeted genome modifications using integrase-deficient lentiviral vectors*. Mol.Ther., 2007. 15(12): pp. 2107–2113.

[37] Joglekar, A.V., et al., *Integrase-defective lentiviral vectors as a delivery platform for targeted modification of adenosine deaminase locus*. Mol Ther, 2013. 21(9): pp. 1705–1717.

[38] Provasi, E., et al., *Editing T cell specificity towards leukemia by zinc finger nucleases and lentiviral gene transfer*. Nat Med, 2012. 18(5): pp. 807–815.

[39] Yi, G., et al., *CCR5 Gene Editing of Resting CD4(+) T Cells by Transient ZFN Expression From HIV Envelope Pseudotyped Nonintegrating Lentivirus Confers HIV-1 Resistance in Humanized Mice.* Mol Ther Nucleic Acids, 2014. 3: p. e198.

[40] Gori, J.L., et al., *Delivery and specificity of CRISPR-Cas9 genome editing technologies for human gene therapy.* Hum Gene Ther, 2015. 26(7): pp. 443–451.

[41] Mock, U., et al., *Novel lentiviral vectors with mutated reverse transcriptase for mRNA delivery of TALE nucleases.* Sci Rep, 2014. 4: p. 6409.

[42] Cai, Y., Bak, R.O., and Mikkelsen, J.G., *Targeted genome editing by lentiviral protein transduction of zinc-finger and TAL-effector nucleases.* Elife, 2014. 3: p. e01911.

[43] Genovese, P., et al., *Targeted genome editing in human repopulating haematopoietic stem cells.* Nature, 2014. 510(7504): pp. 235–240.

[44] Ousterout, D.G., et al., *Correction of dystrophin expression in cells from Duchenne muscular dystrophy patients through genomic excision of exon 51 by zinc finger nucleases.* Mol Ther, 2014. 23(3): pp. 523–532.

[45] Poirot, L., et al., *Multiplex genome-edited T-cell manufacturing platform for "off-the-shelf" adoptive T-cell immunotherapies.* Cancer Res, 2015. 75(18): pp. 3853–3864.

[46] Mock, U., et al., *mRNA transfection of a novel TAL effector nuclease (TALEN) facilitates efficient knockout of HIV co-receptor CCR5.* Nucleic Acids Res, 2015. 43(11): pp. 5560–5571.

[47] Beane, J.D., et al., *Clinical scale zinc finger nuclease-mediated gene editing of PD-1 in tumor infiltrating lymphocytes for the treatment of metastatic melanoma.* Mol Ther, 2015. 23(8): pp. 1380–1390.

[48] Urnov, F.D., et al., *Highly efficient endogenous human gene correction using designed zinc-finger nucleases.* Nature, 2005. 435(7042): pp. 646–651.

[49] Cradick, T.J., et al., *CRISPR/Cas9 systems targeting beta-globin and CCR5 genes have substantial off-target activity.* Nucleic Acids Res, 2010. 41(20): pp. 9584–9592.

[50] Hu, Z., et al., *TALEN-mediated targeting of HPV oncogenes ameliorates HPV-related cervical malignancy.* J Clin Invest, 2014. 125(1): pp. 425–436.

[51] Cong, L., et al., *Multiplex genome engineering using CRISPR/Cas systems.* Science, 2013. 339(6121): pp. 819–823.

[52] Mali, P., et al., *CAS9 transcriptional activators for target specificity screening and paired nickases for cooperative genome engineering.* Nat Biotechnol, 2013. 31(9): pp. 833–838.

[53] Platt, R.J., et al., *CRISPR-Cas9 knockin mice for genome editing and cancer modeling.* Cell, 2014. 159(2): pp. 440–455.

[54] Zuris, J.A., et al., *Cationic lipid-mediated delivery of proteins enables efficient protein-based genome editing in vitro and in vivo.* Nat Biotechnol, 2015. 33(1): pp. 73–80.

[55] Gaj, T., et al., *Targeted gene knockout by direct delivery of zinc-finger nuclease proteins.* Nat Methods, 2012. 9(8): pp. 805–807.

[56] Ru, R., et al., *Targeted genome engineering in human induced pluripotent stem cells by penetrating TALENs.* Cell Regen (Lond), 2013. 2(1): p. 5.

[57] Heitz, F., Morris, M.C., and Divita, G., *Twenty years of cell-penetrating peptides: from molecular mechanisms to therapeutics.* Br J Pharmacol, 2009. 157(2): pp. 195–206.

[58] Liu, J., et al., *Cell-penetrating peptide-mediated delivery of TALEN proteins via bioconjugation for genome engineering.* PLoS One, 2014. 9(1): p. e85755.

[59] Ramakrishna, S., et al., *Gene disruption by cell-penetrating peptide-mediated delivery of Cas9 protein and guide RNA.* Genome Res, 2014. 24(6): pp. 1020–1027.

[60] Cheng, C.J., and Saltzman, W.M., *Enhanced siRNA delivery into cells by exploiting the synergy between targeting ligands and cell-penetrating peptides.* Biomaterials, 2011. 32(26): pp. 6194–6203.

[61] McNeer, N.A., et al., *Nanoparticles that deliver triplex-forming peptide nucleic acid molecules correct F508del CFTR in airway epithelium.* Nat Commun, 2015. 6: p. 6952.

[62] Rogers, F.A., et al., *Site-directed recombination via bifunctional PNA-DNA conjugates.* Proc Natl Acad Sci U S A, 2002. 99(26): pp. 16695–16700.

[63] McNeer, N.A., et al., *Systemic delivery of triplex-forming PNA and donor DNA by nanoparticles mediates site-specific genome editing of human hematopoietic cells in vivo.* Gene Ther, 2013. 20(6): pp. 658–669.

[64] Schleifman, E.B., et al., *Site-specific genome editing in PBMCs with PLGA nanoparticle-delivered PNAs Confers HIV-1 resistance in humanized mice.* Mol Ther Nucleic Acids, 2013. 2: p. e135.

Section 2

Gene Editing and Gene Correction Technologies

Chapter 3

DNA Elements Tetris: A Strategy for Gene Correction

Colette Bastie and Florence Rouleux-Bonnin

Additional information is available at the end of the chapter

http://dx.doi.org/10.5772/62382

Abstract

Transposable elements (TEs) are mobile genetic sequences that are able to move in the genome from one location to another. TEs were first regarded as junk or selfish DNA, as they comprise the largest molecular class within most metazoan genomes having no genomic function. It was necessary to wait until whole genome sequencing to provide new insights about the origin, diversity, and impact of TEs on the genome function. Thus, due to advances in molecular technology, TEs have been shown to create new regulatory sequence networks. Although nowadays most TEs present in the human genome are silenced, particularly DNA transposons, it does not mean that these sequences are dead. In this review, we detail how DNA transposons could be emphasized to create a new tool for gene correction. DNA-based transposon vectors are derived from three models: Sleeping Beauty, piggyBac, and Tol2, which all work via a "cut-and-paste" mechanism where transposase enzyme is alone able to catalyze the transposition process, which means integrating the genes of interest in chromosomal DNA. Limitations and improvements of the systems are discussed, particularly the latest way to target a specific integration site, showing that the DNA transposon-derived system and its engineering, are powerful tools for gene correction.

Keywords: transposon, piggyBac, Sleeping Beauty, gene transfer, Molecular engineering

1. Introduction

1.1. Transposable elements (TEs) in the genome: a brief history from their discovery to their biotechnological use in gene transfer

TEs, also described as "jumping genes," were first discovered in maize by Barbara McClintock in the 1940s. TEs are discrete pieces of DNA that are able to move from one site to another within one genome. This new concept, which suggested that the genome was not a final design but was rather able to evolve, to rearrange, was first met with criticism. However, a large body of evidence has accumulated over the last 60 years not only on the categorization and classification of TEs [1] but also on the understanding of their mechanisms. The ability to accurately identify and classify these sequences is critical to understand their impact on host genomes. Pioneers such as Finnegan [2] classified TEs into two classes based on their mechanism of transposition (**Figure 1**). Class I elements transpose by reverse transcription using an RNA intermediate: they are named retrotransposons. Three kinds of enzyme, RNA polymerase, reverse transcriptase, and integrase, are used for transposition. Class II elements directly transpose from DNA to DNA: they are named DNA transposons and just one enzyme, the transposase, is needed.

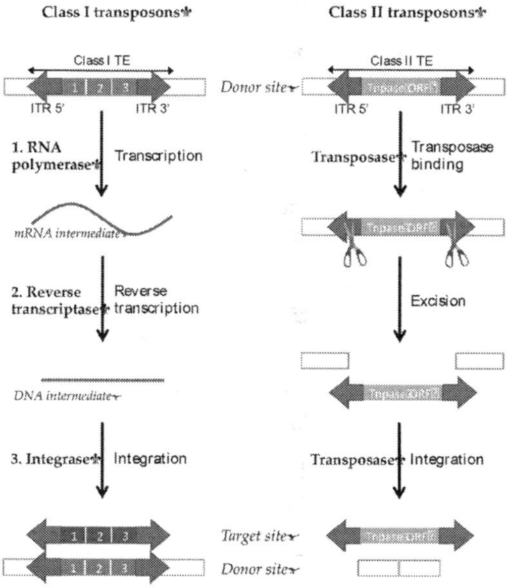

Figure 1. Classes I and II transposable elements (TEs, in green). Class I transposon or RNA transposon: three enzymes are necessary to transpose (1: RNA polymerase, 2: reverse transcriptase and 3: integrase). This mechanism is called "copy-and-paste" and gives rise to two identical copies ; one in the donor site and one in the target site. Class II transposon or DNA transposon: only one enzyme, the transposase, catalyzes the excision and the integration processes. The mechanism is named "cut-and-paste" and translocates the TE element in the target site leaving a free TE donor site. Inverted terminal repeats (ITRs) are drawned in red.

Piégu et al. [1] clearly detailed the necessity to update this classification. TEs are widely distributed in prokaryotic and eukaryotic genomes and represent a variable fraction accounting for 8% in chicken to 85% in maize. After an initial phase of sudden episodic bursts, the invasion step, TEs proliferate and accumulate mutations. Finally, transposition is tolerated by the genome at a reduced rate. Some TE insertions contribute with new genes, exons, or regulator regions. This has been called the exaptation [3] and domestication [4] processes. However, for a significant amount of time, TEs were primarily considered as "junk or selfish DNA" that played no significant role in genome evolution [5]. The modern-day view of TEs is that they can generate genomic instability and reconfigure gene expression networks in both germline and somatic cells. This comprehensive view came with significant advances in sequencing technologies and the development of bioinformatics tools. One of the most unexpected insights is that almost half of our DNA is derived from TEs and 75% of our genome is transcribed (ENCODE project [6]). Therefore, as an integral part of the genome, the dynamic presence of TEs will be a major force to naturally reshape genomes. Several researchers have found examples of concordant timing between bursts of transposition or massive extinction and speciation events. For example, Lynch et al. [7] noticed how transposons transformed the uterine regulatory landscape during the evolution of mammalian pregnancy and Britten [8] reviewed the importance of Alu inserts on brain growth. Thus, TEs are "spam" coming from the dark ages and nowadays a small proportion of retroelements (<0.05%) remains able to transpose in humans [9]. However, no evidence of DNA transposon families was found active in the human genome during the later phase of the primate Radiation, 37 million years ago [10]. The last active DNA transposons were from the hAT superfamily, the Tc1/*mariner*, and the piggyBac families. This suggests that three sources of transposase were silenced at the same evolutionary period. As previously discussed, although transposons have been silenced, it does not mean that they are dead sequences for the genome and they constitute new regulatory networks.

Thus, DNA TEs present distinguishing features, making them attractive as gene transfer tools. Indeed, they are not infectious, as they are able to mobilize DNA in a single genome and are ubiquitous. From the natural architecture of DNA transposons, a secure and easy system has been designed (**Figure 2**).

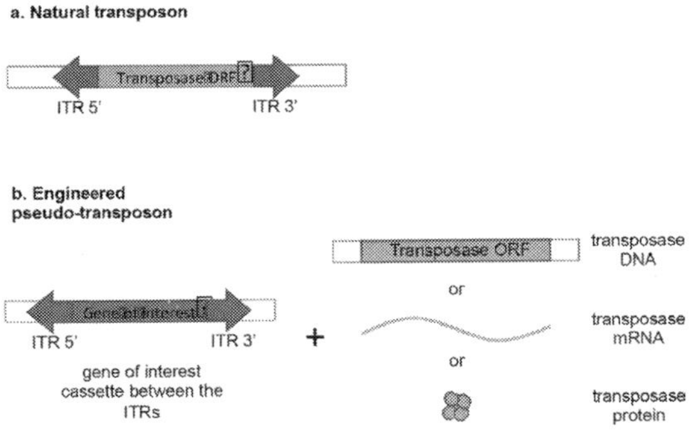

Figure 2. From natural transposon to engineered pseudo-transposon. a) In the natural transposon, the transposase ORF (green rectangle) is delineated by the two ITRs (red arrows). b) In engineered pseudo-transposon, the transposase ORF is replaced by the cassette of the gene of interest. Transposase should therefore be delivered in parallel either in DNA, mRNA or protein form.

Briefly, the transposon is naturally delineated by two inverted terminal repeats (ITRs) framing the unique transposase open reading frame (ORF). The transposase recognizes the ITRs and catalyzes the excision and integration processes (**Figure 1**). After engineering, the transposase ORF is replaced by the gene of interest cassette and the enzyme is brought independently (**Figure 2**). The transposase is then able to integrate any gene of interest, without cross-mobilization between transposon families, as the ITR sequences are highly specific for each transposase. From this global conception of the transposon tool, numerous technological aspects have been explored, finally resulting in an attractive gene integrative system to modify the human genome.

2. Transposon-based strategies

Various transposon-based strategies are available to obtain efficient transgene integration while maintaining safety and cell integrity. First, it depends on the transposase used to govern the efficacy of the integration process. Second, it depends on the way the transposase and the transgene would be delivered. Some use only one plasmid carrying the transposase expression cassette and the transgene construct. Other strategies rely on using one helper molecule carrying the transposase under gene, mRNA, or protein form and one donor plasmid that brings the gene of interest delineated by two ITRs.

2.1. Different types of transposase

For genome engineering, two strategies have been developed: find a transposase in any other species that works in humans or create a new one considering that nowadays no DNA transposons are found active in mammalian genomes. After the identification of efficient transposases for gene correction, their activities have been dissected and optimized.

2.1.1. The three musketeers

For decades, three main transposases have been developed with the aim of gene correction: Sleeping Beauty (SB), piggyBac (PB) and Tol2. In 1997, the SB transposase was artificially reconstructed from partial ancestral copies of a transposase gene identified in salmonid *Salmo* sp. [11]. The Tol2 and piggyBac transposases have been found to be active in their natural host. The piggyBac transposase was isolated from the cabbage looper moth *Trichoplusia ni*, and the developed tool is active in human and mice cells [12]. Tol2 was isolated from the Japanese medaka fish *Oryzias latipes* [13]. It is active in vertebrate cells including zebrafish, chicken, mouse and human.

Following their discovery, various optimizations were carried out to increase their transposition efficiency. The development of the SB100x transposase [14], characterized by a 100-fold greater efficacy than the natural SB, stands as an important step of transposase optimization. Comparatively, in 2011, a hyperactive piggyBac transposase was found with 17- and 9-fold increases in excision and integration, respectively [15], and a codon-optimized PB (mPB) was also developed [16]. Following this, the efficacy of this hyperactive PB (hyPb or 7PB) was compared to SB100x by luciferase *in vivo* expression. Mice injected with m7pB had 10 times greater luciferase expression than those injected with SB100x [17]. Currently, no optimization studies have been carried out On the Tol2 enzyme since it is higly sensitive to molecular engineering [1].

2.1.2. *Transposases confer specific properties to the system*

Naturally, each transposase governs the integration of the pseudo-transposon using their own target site. The integration site for the SB transposon is TA, whereas it is TTAA for the PB transposon and 8-bp target duplication for the Tol2 transposon. After integration, these target sites are duplicates on either side of the newly integrated pseudo-transposon. Besides this specific transposition signature, the SB, PB, and Tol2 transposases confer specific properties to the system, such as cargo size capacity, overproduction inhibition (OPI), and reversibility with or without footprint.

2.1.2.1. *Cargo size capacity*

The distance between ITRs delineates the cassette transgene and defines the cargo size capacity. The more this distance is important, the less the transposase is efficient for excision and integration. However, the constant optimization of the enzymes improved considerably the efficacy of the system.

For now, the SB transposase initially allowed the transposition of only 10-kb transposon [18]. Beyond this size, the transposition rate is abolished. In 2014, Turchiano et al. [19] suggested to change its configuration, permitting the use of SB transposon until 18 kb but with a reduced efficiency. To date, the PB transposon offers the higher cargo size capacity with a natural high activity with 14.3-kb transgenes [12]. The hyPB transposase allows transposition of transgenes up to 100 kb in mouse ES cells [20]. In contrast, Tol2 does not show decrease of transposition efficacy until 10-kb transposon [21], and its activity has been proven until 66 kb [22]. However, few studies have directly compared the transposition efficacy of the transposases in an identical system [23].

Raising cargo size capacity opens new perspectives in gene correction. For example, in muscular dystrophy, disease is induced by the dystrophin mutation. Adding the full-length cDNA of the dystrophin, 11-kb length, has been proven complicated using viral gene transfer. Recently, the full-length dystrophin cDNA has been successfully integrated in mesangioblasts from a dystrophic dog model using the PB transposon tool [24].

2.1.2.2. *Overproduction inhibition*

As previously discussed, the transposase is brought independently to the pseudo-transposon, and the ratio between the enzyme and the pseudo-transposon turns out to be important to establish. On the one hand, transposases act by creating double-stranded breaks so the amount of transposase used must be the lowest possible to avoid genotoxicity. On the other hand, it is necessary to have enough transposase for having high transposition rate. Unexpectedly, increasing the amount of transposase does not result in more transposition activity. Indeed, even if at low level the transposition rate increases with the amount of transposase until a maximum value, it is abolished above. This phenomenon is called OPI and depends on the studied model and the type of transposase [25]. In other cases, the transposition rate is saturated, without decrease, and a plateau is observed. The OPI has been well documented for a long time concerning the SB transposase [26]. However, concerning the PB and Tol2 transposases, the OPI is not as clear. For example, the PB transposase showed an OPI phenomenon in HeLa cells [16], but a stabilization of the activity was demonstrated in HEK293 [27] or mouse ES cells [28]. Similarly, for its Tol2 transposase, OPI or stabilization has been observed [16,21]. The molecular mechanism of this phenomenon is not still clearly established. Numerous hypotheses have been subjected and reviewed in Ref. [25].

2.1.2.3. Integration is reversible

In some conditions, the desired integration needs to be reversed. The transposase could then been readded with the aim of excising the pseudo-transposon from its chromosomal location. The excision of SB pseudo-transposons drives a footprint signature creating a 5-bp insertion [29]. Tol2 transposase excisions have been less investigated, but they could leave a short insertion or deletion [30]. In contrast, PB transposases have the particularity to carry out this excision without leaving a footprint in the genomic sequence. This property has been extensively exploited in induced pluripotent stem cells (iPSC) generation [31–33]. For more security, it is possible to use an engineered PB transposase in which the integration efficacy is abolished while conserving its excision property [34].

2.2. Design of the coupled pseudo-transposon/transposase architecture

Besides the intrinsic particularities of the transposases, the cellular delivery system is crucial. In a first system, called "cis" configuration, only one plasmid carries both the transposase and the gene of interest. The second way, termed "trans" configuration, is based on the principle of separately bringing the gene of interest on one plasmid, "donor" plasmid, and the transposase under a "helper" plasmid or mRNA or protein form.

2.2.1. "Cis" versus "trans" configurations

In the cis configuration, only one plasmid needs to be prepared. This confers easier manipulation and high efficacy, but three drawbacks need to be overcome. First, the pseudo-transposon/transposase ratio is fixed, conferring less flexibility to the system. Second, the plasmid backbone could be integrated as well as, third, the transposase gene. Even if the pseudo-transposon/transposase ratio is fixed, working on promoters has brought flexibility. Indeed, Mikkelsen et al. [35] compared the efficiency of their helper-independent SB vector depending on 11 different promoters used for driving the transposase gene and they observed the OPI phenomenon with the strongest promoter.

In the "trans" configuration, two molecules are used, one carrying the gene of interest and one bringing the transposase either in DNA, RNA, or protein forms. The trans configuration offers naturally more flexibility than the cis one. On the one hand, this approach gives the advantage to modulate the molecular ratio between the transposase and the pseudo-transposon. On the other hand, this approach gives the possibility to introduce several independent pseudo-transposons [36] in their inducible systems. Only one constraint has been detailed: transposases are able to catalyze integration more efficiently with a circular donor plasmid than with a linear one [37].

Type of system	System architecture	Risk of transposase integration	Reference
Cis delivery: one plasmid containing the transposase and the transposon	Conventional cis architecture 	High	[47,48]
	Promoter of the transposase is between the ITRs 	Low and inefficient transposase	[46]
	Transposase and pseudo-transposon share the same polyadenylation signal 	Low and inefficient transposase	[44,49]
Trans delivery: two separate molecules	Conventional trans architecture: two separate plasmids 	Low	[48,50]
	Pseudo-transposon + transposase as mRNA 	Not	[51–54]
	Pseudo-transposon + transposase as protein 	Not	[55–57]

Table 1. Different configurations to deliver transposase and pseudo-transposon and their consequences. Transposase molecules are in green whatever is the molecule type. Pseudo-transposon molecule is drawned in blue. GOI, gene of interest; p(A), polyadenylation signal; Prom, promoter; Tnpase, transposase; ITR, Inverted terminal repeats.

2.2.2. Risks and solutions associated to each strategy

2.2.2.1. Risk of linearized backbone integration

After excision of the gene of interest, the backbone thereby linearized is more prone to be integrated by a nontransposition process [38], whatever the cis or trans configuration used. This undesired integration exposes the problem of the presence of bacterial sequence such as resistance gene or bacterial replication origin. This has been correlated with the amount of transfected transposase [38] and with the size of the transgene [39]. To avoid this, Wilson's team suggested to use a suicide gene in the plasmid backbone, [40] or to select cells expressing green fluorescent protein (GFP) present in the backbone donor plasmid [38]. Other authors suggested using DNA minicircles [41]. Interestingly, they also observed an increased efficacy with DNA minicircles compared to standard plasmid for the same transgene size in several cell lines. However, keeping only the pseudo-transposon as linearized donor plasmid showed no efficacy with SB transposase [42] and a low one with the PB transposase [37].

2.2.2.2. Risk of transposase gene integration

The presence of the transposase gene within the plasmid generates risk of its own integration and *per se* a risk of sustained transposase expression. The consequence could be saltatory remobilization of the integrated transgene [43]. To limit the effect of sustained transposase expression, a self-inactivated transposase gene has been obtained by including either the promoter [44,45] or the polyadenylation signal [46] between the ITRs (**Table 1**). Indeed, in primary human T cells, authors identified an active SB transposase ORF only in one clone out of 94, but a bulk analysis showed up to 0.047 transposase copy integrated per cell [50]. This still has not been evaluated for the PB and Tol2 transposases. Nevertheless, it is possible to completely abolish its integration by introducing transposase under mRNA or protein form (**Table 1**) [51]. mRNA or protein forms allow a one-shot transposition process, thanks to the time-restricted transposase expression.

For example, mRNA transposase expression peaked at 18 h after transfection [58]. Galla et al. [52] demonstrated less cell mortality with the mRNA transposase than an integrative form. Bire et al. [51] showed that the mRNA transposase gave less double-stranded break formation and less copy transgene integration. Moreover, no integrations of the transposase mRNA have been highlighted [51]. These considerations have been confirmed *in vivo* [53], as detailed in the end of this chapter.

Using the protein transposase offers also a short window of expression. Cai et al. [55] recently used the transposase protein associated with viral polyprotein. They observed a high number of transgene expressing cells, with a few number of integrated transgene copies per genome. Aiming to limit viral particle uses, recombinant transposase protein was fused with the cell penetrating peptide (CPP) [56] or transposase was delivered with a free CPP [57]. For now, no *in vivo* evaluations have been found in the bibliographic database.

3. Editing the genome: the final step after a long journey through the cell

Genome editing includes all methods aimed to modify the genome by introducing new DNA sequences or by correcting existing genomic sequences. The journey begins with the ability to enter into the cell, evade the immune response, and, after crossing the nuclear barrier, integrate the gene of interest into the DNA genome.

3.1. Cross the cellular membrane and escape immune response

As free DNA delivery did not show efficient results, both transposase and pseudo-transposon need to be driven into the cell using different gene delivery strategies, either using a carrier (viral particles or chemical agent) or using a physical method. According to the method selected, it is important to consider all parameters of cellular defense against the entry of the foreign DNA.

3.1.1. *Viral hybrid systems*

The viral-transposon hybrid systems take advantage of the natural properties of viral proteins to enter into the cell. For example, as early as 2006, a hybrid HSV amplicon-SB transposase vector was used in a central nervous system development study [59]. Since that time, several studies have been developed on hybrid transposase systems (reviewed in Refs. [60,61]) that use adenovirus [62–64], adeno-associated virus [65], baculovirus [66], or nonintegrative lentivirus [67,68] particles.

3.1.2. *Chemical agents*

Chemical agents have been developed with the aim of condensating DNA and thereby avoiding any viral derived systems. However, it turns out to be more controversial than expected with respect to the immune escape [69]. Indeed, these nanovehicles enter into the cell essentially *via* the endosomal pathway [70,71] and therefore expose foreign DNA to the endosomal Toll-like receptors. Among all available chemical carriers, the polyethylenimine (PEI) polymers appear to be the most used in transposon systems. Indeed, the PEI improve endosomal escape through the "proton sponge" mechanism. For example, in 2009, Kang et al. [72] used the PB transposase-based system with the PEI as a transfection reagent for ovarian cancer treatment in a mouse model. Further examples have been realized both *in vitro* [73] and *in vivo* [36,74].

3.1.3. *Physical gene transfer*

Finally, plasmid DNA could be driven by physical methods. In this case, the plasmid traffic does not go through the endosome and thereby escapes Toll receptors. One such method, electroporation, turned out to be highly efficient to transfect otherwise hard to transfect cells such as dendritic cells and human hematopoietic or embryonic cells [75–77]. Depending on the cell type used, the results may be controversial. Ley et al. [73] compared transposition efficiency in PEI-transfected versus electroporated mesoangioblasts and were not able to obtain efficient long-term expression in muscle after *in vivo* electroporation.

Other physicals methods have therefore been developed. For example, ultrasound targeted microbubble destruction (UTMD) results in pore formation on the cell membrane after ultrasonic waves application. Recently, two *in vivo* studies have been carried out with clinical perspectives [78,79]. In parallel to UTMD, the hydrodynamic (HD) injection has been applied to transfer the clotting factor VIII [80]. However, they are proinflammatory consequences inducing a lack of transgene expression. To circumvent this drawback, Doherty et al. [81] suggested to induce transient transgene repression, thereby preventing the priming of transgene-specific T cells.

3.2. Cross the nuclear barrier and transgene integration

3.2.1. The transposase is driven to the nucleus

For an efficient transposition, the transposase needs to be localized into the nucleus at the same time as the pseudo-transposon DNA.

The transposases contain a nuclear localization signal, driving them to the nucleus [82]. An engineered PB transposase have been developed for increasing its localization within the nucleoli by adding a nucleolus-predominant (NP) signal peptide from HIV-1 TAT protein [83]. With this NP-mPB, a three- to fourfold increase in PB transposition rate, in both murine and human cells, was observed.

From the pseudo-transposon point of view, its nuclear targeting is also essential. Thus, DNA nuclear targeting sequences (DTS) might be added to the plasmid backbone. These DTS consist, for example, to a 72-bp sequence from the SV40 enhancer and act as a sequence driver [84].

3.2.2. Integration profile of the gene of interest

All transposon systems have less integration bias than viruses, as previously described [85–88]. However, it is important to note that there are some differences within transposon systems [89]. The SB transposase is known to allow the more random integration [90], with approximately 35% into RefSeq sequences. It has been notified that the SB transposition has an affinity for the heterochromatin topology [91]. In contrast, the Tol2 and PB transposases are not considered to allow random integration. Indeed, the PB transposase shows a bias towards integration of the transgene into CpG islands and transcriptional start site, with approximately 49% into RefSeq sequences [16,27], and the Tol2 transposase presents a strong bias for the intergenic regions [92].

Interestingly, this global integration profile could be affected by various parameters, such as the transposase variant [93] or the cell type [94].

In addition, it is important to note that, for now, studies have been essentially established in *in vitro* models and no predictions could be drawn regarding the *in vivo* integration profile. Indeed, after *in vivo* UTMD transfection, the pseudo-transposon showed a significant bias of transgene integration into chromosome 14 [49], but no bias was observed in their *in vitro* control.

4. Side effect of the transgene integration system

The newly integrated foreign DNA is considered as an invader by the cell. This leads to postintegrative transgene silencing. Conversly, the transgene copy might also influence surrounding sequences according to the integration site. To conter these mutual side-effects numerous strategies have been developed.

4.1. Communication mechanisms between the transgene and the genome

During their evolution, transposons have been made extinct by at least chromatin condensation and by RNA interference (RNAi) induction.

The transcriptional regulation includes DNA CpG methylation and histone modifications. It has been confirmed that the transgene expression could be restored by a demethylating agent such as 5-aza-2′-deoxycytidine or by a histone deacetylase inhibitor such as trichostatin A [95]. However, it is easier to avoid the induction of upstream gene silencing. To this end, working with a methylated pseudo-transposon plasmid unlike an unmethylated one showed more transposition rate with the SB transposase [96]. Curiously, when the SB, PB, and Tol2 transposase systems are directly compared, the integrated transgene is less silenced if integrated by the PB transposase [97].

The role of RNAi in posttranscriptional silencing of exogenous DNA transposons remains unclear. One study demonstrated that, in the absence of an efficient cellular RNAi system, by establishing p19 protein knockdown cells, the number of colonies is increased [98]. Nonetheless, the mechanism is still not elucidated.

Besides the host-to-transgene effect, a transgene-to-host effect, driving perturbations in sequences surrounding the transgene by DNA methylation modulation, has been highlighted [99]. A further study investigated the expression levels of host genes neighboring the SB transposon and underlined variations depending on the chromosomal location of the transgene [100]. Therefore, solutions allowing a complete isolation of the transgene should be developed.

4.2. Overcoming the host regulation for a sustained expression

In gene correction, maintaining the expression level of the transgene and limiting host genome perturbations are crucial for having an efficient therapeutic effect.

4.2.1. Matrix attachment region (MAR)

The human MAR elements are natural elements of the eukaryotic genome, which mediate the structural organization of the chromatin domains. When included in a transposon plasmid, they do not affect the number of transposed transgene copies but rather increase the transgene expression per integrated copy [101]. Moreover, when the MAR element is included in the transposase vector, an increased transposition efficacy has been observed [102].

4.2.2. Insulators

Insulators are short DNA sequences naturally present in the genome and act as genetic boundary elements. In a recent study, four different insulators (cHS4, D4Z4, CTCF, and CTF/NF1) were compared and showed that D4Z4 and CTF/NF1 had insulator functions when combined with transposition [51]. The protective effect of the cHS4 insulator has been demonstrated by a strong diminution of the activation of a nearby promoter [103] and by a prolonged fluorescent marker expression [104,105]. Some equivalent studies corroborated this role in clinically relevant cells as well as primary hematopoietic $CD34^+$ cells [106]. Moreover, cHS4 insulators abolished the RNAi pathway effects regulating transposon-derived transgene expression by epigenetic silencing [98]. Nevertheless, for an optimal boundarie effect of insulators, it is necessary to consider the model used. Indeed, the size of the pseudo-transposon increased by the insulator or steric hindrance of transposase action [103] could also influence the transgene expression..

5. Going further

For many years, researchers have provided elements for a better understanding of their mechanism and have given solutions for the optimal use of these systems. Here, we recall promising leads for further work in this area: targeting a specific site within the genome and targeting a specific tissue at the body scale.

5.1. Targeting a specific site within the genome

Replacing a defective gene or introducing a gene of interest into a completely safe, predetermined, specific genomic site is the ideal approach for gene correction. This potential locus could be defined by numerous criteria determined by its position from gene, miRNA, transcription unit, or ultraconserved region. All of these aspects have been recently reviewed [107].

5.1.1. Transposon targeting strategies

The SB, PB, and Tol2 transposases have short integration target sites: TA, TTAA, and 8-bp sequences, respectively. Thus, transposon-derived systems should be optimized by combining the transposase to a system able to target a specific DNA sequence, such as a DNA-binding domain (DBD). The first strategy uses a fusion protein containing both the transposase and a DBD. In the second method, a fusion protein is constructed between a DBD and a protein, which is able to specifically recruit the transposase. To date, only one protein is known to be able to interact with the SB transposase, which is named N-57 [108]. Finally, another solution is based on a fusion protein between two DBD, one recognizing a genomic sequence and one specific to a sequence inserted within the pseudo-transposon plasmid. Few parameters of this third approach have been explored in a mammalian model [108]. Considerations of these three strategies have been recently reviewed [109], and we herein detail only chimeric transposases.

The proof-of-concept has been demonstrated by studying intraplasmic integration using the PB transposase fused to the Gal4 domain [110]. However, the system revealed to be more restrictive than expected both in the conservation of the transposition activity and the ability to restrict integration in the targeted locus. Therefore, the transposition activity might be affected by the DBD fusion. Indeed, the DBD Gal4 (a zinc finger domain, ZF) has been tested in fusion to the Tol2, SB11, and PB transposases. The number of chromosomal integrations of the transposon is abolished with Gal4-Tol2 and Gal4-SB11, but no loss of efficiency was observed for the Gal4-PB transposase [111]. Some studies have been carry out to analyze the parameters of this loss of activity, such as the sequence surrounding the targeted site [108], the orientation of the fusion [112], or the choice of the linker [113]. The DBD type has also been evaluated in their ability to avoid off-target integration. With the Gal4-PB transposase, transposition occurred at 23% within 0.8 kb of Gal4 site compared to 5% for the native transposase [114]. However, for improvement of the targeting, artificial ZFs have been created by assembling six ZF domains to create a polydactyl protein capable of targeting a unique sequence of 18 bp [115]. For example, the sequence targeting with these artificial ZF allowed 44.3% of integration events near the CHK2-ZF site [116]. Comparatively, when the Sp1 ZF is fused with the PB transposase, which preferentially binds the CG-rich motif, the integration increased near the CpG islands (25.7% versus 10.5% with the native PB transposase) but without modification regarding the integration into the RefSeq genes [117].

5.1.2. Other systems allowing a targeting integration

In 2011, the discovery of the CRISPR/Cas9 system revolutionized the gene transfer because of its ability to drive the transgene in its physiological site, but no studies directly compared the efficiency of both transposon and CRISPR/Cas9-based systems. It has been supposed that this system arises from casposon in the evolutionary tree. Casposons are mobile cryptic sequences present in Achaea and bacteria, and two independent studies described this superfamily of mobile elements by linking transposon and CRISPR/Cas systems [118,119].

Recently, a combinatory approach was developed, in which the correction is realized gene by gene (CRISPR/Cas9 role) and temporarily needed sequences are removed from the genome (transposase role). This method has been applied for gene correction of β-thalassemia [120] and to create iPSC with deletion into the CCR5 gene [121].

5.2. Targeting a specific tissue at the organism scale

For *in vivo* application of gene correction, it is important to express the transgene of interest only in the organ, tissue, or cell types in which the transgene expression is required. The design of the transgene vector is essential and might contain specific elements such as tissue-specific promoter or regulatory sequences. The second option is to deliver the system only in the specific cells.

5.2.1. Design of the transgene vector for in vivo applications

In the ideal gene transfer, the transgene is expressed in the same conditions, as it is in physiological conditions. Indeed, overexpression of the transgene or expression in a nontarget cell could improve cytotoxicity, induce its clearance by the immune system, and increase its gene silencing (reviewed in Ref. [122]). With this aim, vectors have been designed in such a way as promoters or regulatory sequences are chosen for restricting the expression of the gene of interest only in the cells of interest. Tissue-specific promoters control gene expression in a tissue-dependent manner or according to the development stage of the cells. In plasmid design, several approaches are available such as using a promoter regulating an endogenous gene expressed in one type of cell (minimal promoter) or combining numerous enhancers to a minimal promoter.

In the first case, the transposon is under a native promoter. For example, endothelin-1 [123] allows a decreased GFP expression in a nonendothelial cell line while maintaining the expression level in endothelial cell lines. When the targeted cell type is the final point of a differentiation lineage, it seems essential to have the expression of the therapeutic protein only in the differentiated state, such as promoters capable of restricting β-globin expression in differentiated erythroid cells from transfected proerythroid cells [124]. In cancer therapy, a study based on the SB transposition showed that the HSV-TK transgene driven by a telomerase reverse transcriptase promoter increased death rate in cancer cell lines compared to fibroblast cell lines [125].

The second approach is based on constructions containing a minimal promoter with specific enhancers. For example, the SB transposon system has been used for the introduction of the telomerase gene driven by a combination of the transthyretin (TTR) gene promoter/enhancer, the human alcohol dehydrogenase gene promoter, and the SV40 enhancer [126]. The authors observed an induced transcriptional activity only in hepatocytes. In an *in vivo* study, the authors developed a TTR minimal promoter coupled to a hepatocyte-specific cis-regulatory module, driving the clotting factor IX for correction of hemophilia B [127]. This promoter has also been combined with a PB transposon-mediated gene transfer and confirmed the high efficiency of the transgene construct [128].

5.2.2. Limiting the ectopic integrations by tissue targeting

For improvement of tissue targeting, two major routes have been developed, either administration of *ex vivo* premodified cells of interest or direct delivery of the integrative system, containing the transgene, to the whole organism.

5.2.2.1. Administration route for ex vivo modified cells

The delivery of premodified cells to a patient was extensively carried out in adoptive cell transfer of immune cells expressing an artificial T-cell receptor (TCR) designed to target an antigen. Briefly, T cells are removed from a patient and transformed to express the artificial TCR (also named chimeric antigen receptor or CAR). After amplification, modified T cells are intravenously readministrated to the organism. In the field of transposon technology, this approach has been used in several applications. For example, a human epidermal growth factor receptor 2-specific CAR was introduced into cytotoxic T cells, thanks to the PB transposase [129]. More recently, T lymphocytes were modified to express the CD19-CAR transgene, and after 7 days of coculture, CAR T cells eradicated all $CD19^{+}$ tumor cells *in vitro* [130]. In lower proportions, the Tol2 transposase has also been used for the integration of a CD19-CAR into T cells [131]. However, production of CD19-CAR T cells usually uses SB transposase and clinical trials are currently under investigation [132]. The authors detailed their protocol for manufacturing clinical-grade CD19-specific T cells [76].

It is also possible to reimplant modified cells *in situ* after their encapsulation. In this aim, Fjord-Larsen et al. [133] developed a model in which a new clinical-grade cell line expresses a high level of neural growth factor after striatum implantation.

The administration of already modified cells increases the security of the transfer system. However, applications are, for now, restricted to cells easy to collect and reimplant to a patient. For less accessible tissue or organs, targeting methods are more often driven by a direct administration of the transgene.

5.2.2.2. Administration route for transposon DNA system

The administration of the therapeutic gene, associated with the transposase, needs a delivery method able to drive them into the organ or tissue of interest. To this end, two strategies have been developed. The first one takes advantages of specific administration route properties, whereas the second one uses vehicles expressing receptors capable of specific recognition of the targeting tissue.

It has been demonstrated that all gene delivery methods do not present an equal distribution in the different organs. For example, the HD injection is known to target the liver at 95%, as detailed by Bell et al. [134]. In agreement, Herweijer and Wolff [135] showed that transgene expression was also found in others organs such as the heart, spleen, and kidneys at levels approximately 100-fold lower than in the liver. This liver targeting way has been applied in gene correction, and in 2007, Aronovich et al. showed a model of correction of mucopolysaccharidosis mice by SB-mediated transgene α-L-iduridase (IDUA) transposition [136]. They mentioned a persistent expression of IDUA in plasma for almost 10 weeks after injection. In cancer therapy, liver metastasis of colorectal cancer was reduced after antiangiogenic genes were integrated by the SB transposase [137].

As a complement, the DNA transposon could also been administrated after complexation to a targeting vehicle. After an intravenous administration, Kren et al. [47] highlighted a hepatocyte-specific integration of the transgene when condensated with coated nanocapsules. Comparatively, the transgene complexed to the PEI showed an expression in the lung, not observed after HD injection [138]. More specifically, within the lung, the polyplexes are addressed into pneumocytes and no transgene expression was detected within the conducting airways [139].

Coupling specific administration route and nanocapsules is the future way. In this aim, the UTMD gene delivery method allows mediating the site-specific delivery of transposons. Briefly, the transgene is intravenously injected and cell penetration occurs at the targeted organ by acoustic cavitation [49]. This approach has been used for the transposition of the Nkx2.2 transcriptional factor to the pancreas by the PB system [78] or for the transposition of the thymosine β4 gene, or the glucagon-like peptide-1 one, to the heart [79,140].

In gene correction, targeting the tissue of interest is essential for reflecting physiological conditions. Compared to viral transduction, the transposon systems are more customizable and numerous possibilities are available for users. Depending on the tissue to target, it is possible to play at the same time on the promoter, the administration route, and the presence of targeting molecules.

6. Therapy applications of transposase tools

Some technological aspects previously discussed offer a suitable transposon toolbox to gene correction. Transposon-based systems allow first the transgene integration in a large range of clinically relevant target cells, including hematopoietic stem cells [141], mesenchymal stromal cells [142], iPSC [143], and lymphoid T cells [131]. Transposon-mediated correction could therefore be used in a large-scale application, such as treatment of inherited disorders, cancer, and tissue degeneration (**Table 2**).

Disease	Transgene	Tnpase	Animal model or cell type	Reference
Inherited disorders				
Hemophilia A	*FVIII*	SB/PB	Hemophilic A mice	[47,80,144–146]
Hemophilia B	*FIX*	SB/PB	Hemophilic B mice or dogs	[42,62,128,147]
Huntington's disease	*siRNA-htt*	SB	Human cell lines	[148]
Duchene muscular dystrophy	*Dystrophin*	PB	Dog dystrophic mesoangioblast	[24]
Tyrosinemia type I	*Fah*	SB	FAH-deficient mice	[149–151]
Sickle cell disease	*HO-1* or *IHK*	SB	Mice	[152,153]
Mucopolysaccharidosis type I (MPS I)	*hIDUA* or *hGUSB*	SB	MPS I NOD/SCID mice	[136,154,155]
α-Antitrypsin deficiency	*hAAT*	PB	Mice and iPSC	[93,156]
Fanconi anemia type C	*FA-C*	SB	Human lymphoblastoid cells	[157]
Crigler-Najjar syndrome type 1	*hUGT1A1*	SB	Hepa1 cell line and gunn rats	[158]
Junctional epidermolysis bullosa (JEB)	*LAMB3*	SB	Epidermal holoclones from JEB patients	[159]
Vaccination				
Immunization against non-self protein	*eGFP*	PB	Mice	[160]
Regenerative medicine				
iPSC generation	*SOX2, OCT4, KLF4,* and *c-MYC*	PB	Fibroblasts, melanoma cells, HDDPC	[32,143,161–164]
Diabetes	*Nkx2.2* or *insulin*	PB/SB	STZ-rat pancreas	[78,165]
Retinal degeneration	*PEDF*	SB	IPE and RPE cells	[166]
Acute myocardial infraction	*TB4* or *GLP1*	PB	Rat heart	[79,140]
Cancerology				

Disease	Transgene	Tnpase	Animal model or cell type	Reference
Angiogenesis-dependent tumors	*sFlt-1* or *statin-AE*	SB	Tumor engrafts in mice	[66,137,167]
Cervical cancer	*HSV-tk*	PB	Cervical cancer xenografts	[72]
Ovarian adenocarcinoma	*HSV-tk*	PB	Cell line	[168]
Melanoma	*TRAIL and IFNγ*	PB	ADSC	[169]
Adoptive T-cell therapy	*CD19-CAR*, *HER2-CAR*, or *IL-11-CAR*	SB, PB or Tol2	Human T cells and clinical trials	[40,77,129,130,132, 170–179]
Pulmonary and other diseases				
Pulmonary fibrosis	*miR-29* or *hIDO*	SB	Bleomycin-induced pulmonary fibrosis mice or rats	[180,181]
Pulmonary hypertension	*eNOS*	SB	Monocrotaline-induced pulmonary hypertension rats	[182]
Acute cellular injury	*hTERT*	SB	Primary hepatocytes	[126]
Unilateral ureteral obstruction	*IGF-1R*	PB	Mice	[183]

Tnpase, transposase ; ADSC, adipose-derived mesenchymal stem cells; eGFP, enhanced GFP; eNOS, endothelial nitric oxide synthase; FA-C, Fanconi anemia complementation group C; Fah, fumaryl-acetoacetate hydrolase; FVIII, clotting factor, factor VIII; hAAT, human α1-antitrypsin; HDDPC, primary human deciduous tooth dental pulp cells; HER2, human epidermal growth factor receptor 2; hGUSB, β-glucoronidase; hIDO, human indoleamine-2,3-dioxygenase; hIDUA, human α-L-iduronidase; HO-1, heme oxygenase-1; HSV-tk, herpes simplex virus thymidine kinase; hTERT, human telomerase reverse transcriptase; htt, huntingtin; hUGT1A1, human uridine diphosphoglucuronate glucuronosyltransferase 1A1; IFNγ, interferon γ; IGF-1R, insulin-like growth factor-1 receptor; IHK, antisickling globin; IL-11, interleukin-11; IPE, iris epithelial cells; KLF4, Krüppel-like factor 4; LAMB3, laminin B3 subunit of laminin 5; Nkx2.2, NK-type homeodomain transcription factor; OCT4, octamer-binding transcription factor 4; PEDF, pigment epithelium-derived factor; RPE, pigment epithelial cells; sFlt-1, soluble fms-like tyrosine kinase-1; SOX2, SRY (sex-determining region Y) box 2; statin-AE, angiostatin-endostatin fusion gene; STZ, streptozotocin; TB4, thymosin β4; Tnpase, transposase; TRAIL, TNF-related apoptosis-inducing ligand.

Table 2. Application fields of transposon-based gene correction.

7. Conclusion

Transposons have naturally drawn genomes since the first forms of life. Scientists have taken advantage of their properties with the aim of constantly updating the safety of this nonviral tool for gene transfer. With the other integrative systems derived from casposons, such as

CRISPR/Cas9, we dispose of complementary tools for reshaping the genome. Latest discoveries have open new horizons, but a long road is still ahead.

Acknowledgements

This work has been supported by La Ligue Contre le Cancer.

Author details

Colette Bastie and Florence Rouleux-Bonnin*

*Address all correspondence to: florence.bonnin@univ-tours.fr

LNOX GICC UMR CNRS 7292, Department of Medicine, Bâtiment Dutrochet, 10 boulevard Tonnellé, 37032 TOURS, France

References

[1] Piégu B, Bire S, Arensburger P, Bigot Y. A survey of transposable element classification systems—A call for a fundamental update to meet the challenge of their diversity and complexity. Mol Phylogenet Evol. 2015;86:90–109.

[2] Finnegan DJ. Eukaryotic transposable elements and genome evolution. Trends Genet. 1989;5:103–7.

[3] de Souza FSJ, Franchini LF, Rubinstein M. Exaptation of transposable elements into novel cis-regulatory elements: is the evidence always strong? Mol Biol Evol. 2013;30:1239–51.

[4] Alzohairy AM, Gyulai G, Jansen RK, Bahieldin A. Transposable elements domesticated and neofunctionalized by eukaryotic genomes. Plasmid. 2013;69:1–15.

[5] Doolittle WF, Sapienza C. Selfish genes, the phenotype paradigm and genome evolution. Nature. 1980;284:601–3.

[6] Birney E, Stamatoyannopoulos JA, Dutta A, GuigóRR, Gingeras TR, Margulies EH, et al. Identification and analysis of functional elements in 1% of the human genome by the ENCODE pilot project. Nature. 2007;447:799–816.

[7] Lynch VJ, Leclerc RD, May G, Wagner GP. Transposon-mediated rewiring of gene regulatory networks contributed to the evolution of pregnancy in mammals. Nat Genet. 2011;43:1154–9.

[8] Britten RJ. Transposable element insertions have strongly affected human evolution. Proc Natl Acad Sci. 2010;107:19945–8.

[9] Muotri AR, Marchetto MCN, Coufal NG, Gage FH. The necessary junk: new functions for transposable elements. Hum Mol Genet. 2007;16:R159–67.

[10] Pace JKI, Feschotte C. The evolutionary history of human DNA transposons: evidence for intense activity in the primate lineage. Genome Res. 2007;17:422–32.

[11] Ivics Z, Hackett PB, Plasterk RH, Izsvák Z. Molecular reconstruction of Sleeping Beauty, a Tc1-like transposon from fish, and its transposition in human cells. Cell. 1997;91:501–10.

[12] Ding S, Wu X, Li G, Han M, Zhuang Y, Xu T. Efficient transposition of the piggyBac (PB) transposon in mammalian cells and mice. Cell. 2005;122:473–83.

[13] Kawakami K, Noda T. Transposition of the Tol2 element, an Ac-like element from the Japanese medaka fish *Oryzias latipes,* in mouse embryonic stem cells. Genetics. 2004;166:895–9.

[14] Mátés L, Chuah MKL, Belay E, Jerchow B, Manoj N, Acosta-Sanchez A, et al. Molecular evolution of a novel hyperactive Sleeping Beauty transposase enables robust stable gene transfer in vertebrates. Nat Genet. 2009;41:753–61.

[15] Yusa K, Zhou L, Li MA, Bradley A, Craig NL. A hyperactive piggyBac transposase for mammalian applications. Proc Natl Acad Sci. 2011;108:1531–6.

[16] Grabundzija I, Irgang M, Mátés L, Belay E, Matrai J, Gogol-Döring A, et al. Comparative analysis of transposable element vector systems in human cells. Mol Ther. 2010;18:1200–9.

[17] Doherty JE, Huye LE, Yusa K, Zhou L, Craig NL, Wilson MH. Hyperactive piggyBac gene transfer in human cells and *in vivo*. Hum Gene Ther. 2012;23:311–20.

[18] Zayed H, Izsvák Z, Walisko O, Ivics Z. Development of hyperactive Sleeping Beauty transposon vectors by mutational analysis. Mol Ther. 2004;9:292–304.

[19] Turchiano G, Latella MC, Gogol-Döring A, Cattoglio C, Mavilio F, Izsvák Z, et al. Genomic analysis of Sleeping Beauty transposon integration in human somatic cells. PLoS One. 2014;9:e112712.

[20] Li MA, Turner DJ, Ning Z, Yusa K, Liang Q, Eckert S, et al. Mobilization of giant piggyBac transposons in the mouse genome. Nucleic Acids Res. 2011;39:e148.

[21] Balciunas D, Wangensteen KJ, Wilber A, Bell J, Geurts A, Sivasubbu S, et al. Harnessing a high cargo-capacity transposon for genetic applications in vertebrates. PLoS Genet. 2006;2:e169.

[22] Suster ML, Sumiyama K, Kawakami K. Transposon-mediated BAC transgenesis in zebrafish and mice. BMC Genomics. 2009;10:477.

[23] Rostovskaya M, Fu J, Obst M, Baer I, Weidlich S, Wang H, et al. Transposon-mediated BAC transgenesis in human ES cells. Nucleic Acids Res. 2012;40:e150.

[24] Loperfido M, Jarmin S, Dastidar S, Di Matteo M, Perini I, Moore M, et al. piggyBac transposons expressing full-length human dystrophin enable genetic correction of dystrophic mesoangioblasts. Nucleic Acids Res. 2015 Dec 17;44(2):744–60.

[25] Bire S, Casteret S, Arnaoty A, Piégu B, Lecomte T, Bigot Y. Transposase concentration controls transposition activity: myth or reality? Gene. 2013;530:165–71.

[26] Lohe AR, Hart DL. Autoregulation of mariner transposase activity by overproduction and dominant-negative complementation. Mol Biol Evol. 1996;13:549–55.

[27] Wilson MH, Coates CJ, George AL. PiggyBac transposon-mediated gene transfer in human cells. Mol Ther. 2007;15:139–45.

[28] Cadiñanos J, Bradley A. Generation of an inducible and optimized piggyBac transposon system. Nucleic Acids Res. 2007;35:e87.

[29] Liu G, Aronovich EL, Cui Z, Whitley CB, Hackett PB. Excision of Sleeping Beauty transposons: parameters and applications to gene therapy. J Gene Med. 2004;6:574–83.

[30] Huang P, Xu L, Liang W, Tam CI, Zhang Y, Qi F, et al. Genomic deletion induced by Tol2 transposon excision in zebrafish. Nucleic Acids Res. 2013;41:e36.

[31] Igawa K, Kokubu C, Yusa K, Horie K, Yoshimura Y, Yamauchi K, et al. Removal of reprogramming transgenes improves the tissue reconstitution potential of keratinocytes generated from human induced pluripotent stem cells. Stem Cells Transl Med. 2014;3:992–1001.

[32] Talluri TR, Kumar D, Glage S, Garrels W, Ivics Z, Debowski K, et al. Non-viral reprogramming of fibroblasts into induced pluripotent stem cells by Sleeping Beauty and piggyBac transposons. Biochem Biophys Res Commun. 2014;450:581–7.

[33] Hu K. All roads lead to induced pluripotent stem cells: the technologies of iPSC generation. Stem Cells Dev. 2014;23:1285–300.

[34] Li X, Burnight ER, Cooney AL, Malani N, Brady T, Sander JD, et al. piggyBac transposase tools for genome engineering. Proc Natl Acad Sci. 2013;110:E2279–87.

[35] Mikkelsen JG, Yant SR, Meuse L, Huang Z, Xu H, Kay MA, et al. Helper-independent Sleeping Beauty transposon-transposase vectors for efficient nonviral gene delivery and persistent gene expression *in vivo*. Mol Ther. 2003;8:654–65.

[36] Saridey SK, Liu L, Doherty JE, Kaja A, Galvan DL, Fletcher BS, et al. PiggyBac transposon-based inducible gene expression *in vivo* after somatic cell gene transfer. Mol Ther. 2009;17:2115–20.

[37] Nakanishi H, Higuchi Y, Kawakami S, Yamashita F, Hashida M. Comparison of piggyBac transposition efficiency between linear and circular donor vectors in mammalian cells. J Biotechnol. 2011;154:205–8.

[38] Saha S, Woodard LE, Charron EM, Welch RC, Rooney CM, Wilson MH. Evaluating the potential for undesired genomic effects of the piggyBac transposon system in human cells. Nucleic Acids Res. 2015;43:1770–82.

[39] Nakazawa Y, Saha S, Galvan DL, Huye L, Rollins L, Rooney CM, et al. Evaluation of long-term transgene expression in piggyBac-modified human T lymphocytes. J Immunother. 2013;36:3–10.

[40] Nakazawa Y, Huye LE, Dotti G, Foster AE, Vera JF, Manuri PR, et al. Optimization of the PiggyBac transposon system for the sustained genetic modification of human T lymphocytes. J Immunother. 2009;32:826–36.

[41] Sharma N, Cai Y, Bak RO, Jakobsen MR, Schrøder LD, Mikkelsen JG. Efficient Sleeping Beauty DNA transposition from DNA minicircles. Mol Ther Nucleic Acids. 2013;2:e74.

[42] Yant SR, Ehrhardt A, Mikkelsen JG, Meuse L, Pham T, Kay MA. Transposition from a gutless adeno-transposon vector stabilizes transgene expression *in vivo*. Nat Biotechnol. 2002;20:999–1005.

[43] Li MA, Pettitt SJ, Eckert S, Ning Z, Rice S, Cadinanos J, et al. The piggyBac transposon displays local and distant reintegration preferences and can cause mutations at noncanonical integration sites. Mol Cell Biol. 2013;33:1317–30.

[44] Urschitz J, Kawasumi M, Owens J, Morozumi K, Yamashiro H, Stoytchev I, et al. Helper-independent piggyBac plasmids for gene delivery approaches: strategies for avoiding potential genotoxic effects. Proc Natl Acad Sci. 2010;107:8117–22.

[45] Urschitz J, Moisyadi S. Transpositional transgenesis with piggyBac. Mob Genet Elements. 2013;3:e25167.

[46] Chakraborty S, Ji H, Chen J, Gersbach CA, Leong KW. Vector modifications to eliminate transposase expression following piggyBac-mediated transgenesis. Sci Rep. 2014;4:7403.

[47] Kren BT, Unger GM, Sjeklocha L, Trossen AA, Korman V, Diethelm-Okita BM, et al. Nanocapsule-delivered Sleeping Beauty mediates therapeutic Factor VIII expression in liver sinusoidal endothelial cells of hemophilia A mice. J Clin Invest. 2009;119:2086–99.

[48] Huang X, Wilber AC, Bao L, Tuong D, Tolar J, Orchard PJ, et al. Stable gene transfer and expression in human primary T cells by the Sleeping Beauty transposon system. Blood. 2006;107:483–91.

[49] Anderson CD, Urschitz J, Khemmani M, Owens JB, Moisyadi S, Shohet RV, et al. Ultrasound directs a transposase system for durable hepatic gene delivery in mice. Ultrasound Med Biol. 2013;39:2351–61.

[50] Huang X, Haley K, Wong M, Guo H, Lu C, Wilber A, et al. Unexpectedly high copy number of random integration but low frequency of persistent expression of the Sleeping Beauty transposase following trans delivery in primary human T cells. Hum Gene Ther. 2010;14:1–14.

[51] Bire S, Ley D, Casteret S, Mermod N, Bigot Y, Rouleux-Bonnin F. Optimization of the piggyBac transposon using mRNA and insulators: toward a more reliable gene delivery system. PLoS One. 2013;8:e82559.

[52] Galla M, Schambach A, Falk CS, Maetzig T, Kuehle J, Lange K, et al. Avoiding cytotoxicity of transposases by dose-controlled mRNA delivery. Nucleic Acids Res. 2011;39:7147–60.

[53] Bell JB, Aronovich EL, Schreifels JM, Beadnell TC, Hackett PB. Duration of expression and activity of Sleeping Beauty transposase in mouse liver following hydrodynamic DNA delivery. Mol Ther. 2010;18:1796–802.

[54] Wilber A, Frandsen JL, Geurts JL, Largaespada DA, Hackett PB, McIvor RS. RNA as a source of transposase for Sleeping Beauty-mediated gene insertion and expression in somatic cells and tissues. Mol Ther. 2006;13:625–30.

[55] Cai Y, Bak RO, Krogh LB, Staunstrup NH, Moldt B, Corydon TJ, et al. DNA transposition by protein transduction of the piggyBac transposase from lentiviral Gag precursors. Nucleic Acids Res. 2014;42:e28.

[56] Lee C-Y, Li J-F, Liou J-S, Charng Y-C, Huang Y-W, Lee H-J. A gene delivery system for human cells mediated by both a cell-penetrating peptide and a piggyBac transposase. Biomaterials. 2011;32:6264–76.

[57] Järver P, Fernaeus S, EL-Andaloussi S, Tjörnhammar M-L, Langel Ü. Co-transduction of Sleeping Beauty transposase and donor plasmid via a cell-penetrating peptide: a simple one step method. Int J Pept Res Ther. 2007;14:58–63.

[58] Bire S, Gosset D, Jégot G, Midoux P, Pichon C, Rouleux-Bonnin F. Exogenous mRNA delivery and bioavailability in gene transfer mediated by piggyBac transposition. BMC Biotechnol. 2013;13:75.

[59] Bowers WJ, Mastrangelo MA, Howard DF, Southerland HA, Maguire-Zeiss KA, Federoff HJ. Neuronal precursor-restricted transduction via in utero CNS gene delivery of a novel bipartite HSV amplicon/transposase hybrid vector. Mol Ther. 2006;13:580–8.

[60] Müther N, Noske N, Ehrhardt A. Viral hybrid vectors for somatic integration—are they the better solution? Viruses. 2009;1:1295–324.

[61] Skipper KA, Andersen PR, Sharma N, Mikkelsen JG. DNA transposon-based gene vehicles—scenes from an evolutionary drive. J Biomed Sci. 2013;20:92.

[62] Hausl MA, Zhang W, Müther N, Rauschhuber C, Franck HG, Merricks EP, et al. Hyperactive Sleeping Beauty transposase enables persistent phenotypic correction in mice and a canine model for hemophilia B. Mol Ther. 2010;18:1896–906.

[63] Zhang W, Muck-Hausl M, Wang J, Sun C, Gebbing M, Miskey C, et al. Integration profile and safety of an adenovirus hybrid-vector utilizing hyperactive Sleeping Beauty transposase for somatic integration. PLoS One. 2013;8:e75344.

[64] Smith RP, Riordan JD, Feddersen CR, Dupuy AJ. A hybrid adenoviral vector system achieves efficient long-term gene expression in the liver via piggyBac transposition. Hum Gene Ther. 2015;26:377–85.

[65] Zhang W, Solanki M, Müther N, Ebel M, Wang J, Sun C, et al. Hybrid adeno-associated viral vectors utilizing transposase-mediated somatic integration for stable transgene expression in human cells. PLoS One. 2013;8:e76771.

[66] Luo W-Y, Shih Y-S, Hung C-L, Lo K-W, Chiang C-S, Lo W-H, et al. Development of the hybrid Sleeping Beauty: baculovirus vector for sustained gene expression and cancer therapy. Gene Ther. 2012;19:844–51.

[67] Staunstrup NH, Moldt B, Mátés L, Villesen P, Jakobsen M, Ivics Z, et al. Hybrid lentivirus-transposon vectors with a random integration profile in human cells. Mol Ther. 2009;17:1205–14.

[68] Vink CA, Gaspar HB, Gabriel R, Schmidt M, McIvor RS, Thrasher AJ, et al. Sleeping beauty transposition from nonintegrating lentivirus. Mol Ther. 2009;17:1197–204.

[69] Sakurai H, Kawabata K, Sakurai F, Nakagawa S, Mizuguchi H. Innate immune response induced by gene delivery vectors. Int J Pharm. 2008;354:9–15.

[70] Bieber T, Meissner W, Kostin S, Niemann A, Elsasser H-P. Intracellular route and transcriptional competence of polyethylenimine-DNA complexes. J Control Release. 2002;82:441–54.

[71] Rejman J, Bragonzi A, Conese M. Role of clathrin- and caveolae-mediated endocytosis in gene transfer mediated by lipo- and polyplexes. Mol Ther. 2005;12:468–74.

[72] Kang Y, Yu W, Sun Q, Zhang X, Jiang W, Wu C, et al. High-level transgene expression mediated by the piggyBac transposon enhances transgenic therapeutic effects in cervical cancer xenografts. Oncol Rep. 2010;24:897–907.

[73] Ley D, Van Zwieten R, Puttini S, Iyer P, Cochard A, Mermod N. A PiggyBac-mediated approach for muscle gene transfer or cell therapy. Stem Cell Res. 2014;13:390–403.

[74] Lin E-H, Keramidas M, Rome C, Chiu W-T, Wu C-W, Coll J-L, et al. Lifelong reporter gene imaging in the lungs of mice following polyethyleneimine-mediated Sleeping-Beauty transposon delivery. Biomaterials. 2011;32:1978–85.

[75] Chicaybam L, Sodre AL, Curzio BA, Bonamino MH. An efficient low cost method for gene transfer to T lymphocytes. PLoS One. 2013;8:e60298.

[76] Singh H, Figliola MJ, Dawson MJ, Olivares S, Zhang L, Yang G, et al. Manufacture of clinical-grade CD19-specific T cells stably expressing chimeric antigen receptor using Sleeping Beauty system and artificial antigen presenting cells. PLoS One. 2013;8:e64138.

[77] Singh H, Huls H, Kebriaei P, Cooper LJN. A new approach to gene therapy using Sleeping Beauty to genetically modify clinical-grade T cells to target CD19. Immunol Rev. 2014;257:181–90.

[78] Chen S, Shimoda M, Chen J, Matsumoto S, Grayburn PA, Paul A. Ectopic transgenic expression of NKX2.2 induces differentiation of adult pancreatic progenitors and mediates islet regeneration. Cell Cycle. 2012;11:1544–53.

[79] Chen S, Shimoda M, Chen J, Grayburn PA. Stimulation of adult resident cardiac progenitor cells by durable myocardial expression of thymosin beta 4 with ultrasound-targeted microbubble delivery. Gene Ther. 2013;20:225–33.

[80] Staber JM, Pollpeter MJ, Arensdorf A, Sinn PL, Rutkowski DT, McCray PB. piggyBac-mediated phenotypic correction of factor VIII deficiency. Mol Ther Methods Clin Dev. 2014;1:14042.

[81] Doherty JE, Woodard LE, Bear AS, Foster AE, Wilson MH. An adaptable system for improving transposon-based gene expression *in vivo* via transient transgene repression. FASEB J. 2013;27:3753–62.

[82] Keith JH, Fraser TS, Fraser MJ. Analysis of the piggyBac transposase reveals a functional nuclear targeting signal in the 94 c-terminal residues. BMC Mol Biol. 2008;9:72.

[83] Hong J-B, Chou F-J, Ku AT, Fan H-H, Lee T-L, Huang Y-H, et al. A nucleolus-predominant piggyBac transposase, NP-mPB, mediates elevated transposition efficiency in mammalian cells. PLoS One. 2014;9:e89396.

[84] Miller AM, Dean DA. Tissue-specific and transcription factor-mediated nuclear entry of DNA. Adv Drug Deliv Rev. 2009;61:603–13.

[85] de Jong J, Akhtar W, Badhai J, Rust AG, Rad R, Hilkens J, et al. Chromatin landscapes of retroviral and transposon integration profiles. PLoS Genet. 2014;10:e1004250.

[86] Li X, Ewis H, Hice RH, Malani N, Parker N, Zhou L, et al. A resurrected mammalian hAT transposable element and a closely related insect element are highly active in human cell culture. Proc Natl Acad Sci. 2013;110:E478–87.

[87] Moldt B, Miskey C, Staunstrup NH, Gogol-Döring A, Bak RO, Sharma N, et al. Comparative genomic integration profiling of Sleeping Beauty transposons mobilized with high efficacy from integrase-defective lentiviral vectors in primary human cells. Mol Ther. 2011;19:1499–510.

[88] Field A-C, Vink C, Gabriel R, Al-Subki R, Schmidt M, Goulden N, et al. Comparison of lentiviral and Sleeping Beauty mediated αβ T cell receptor gene transfer. PLoS One. 2013;8:e68201.

[89] Huang X, Guo H, Tammana S, Jung Y-C, Mellgren E, Bassi P, et al. Gene transfer efficiency and genome-wide integration profiling of Sleeping Beauty, Tol2, and piggyBac transposons in human primary T cells. Mol Ther. 2010;18:1803–13.

[90] Yant SR, Wu X, Huang Y, Garrison B, Burgess SM, Kay MA. High-resolution genome-wide mapping of transposon integration in mammals. Mol Cell Biol. 2005;25:2085–94.

[91] Ikeda R, Kokubu C, Yusa K, Keng VW, Horie K, Takeda J. Sleeping beauty transposase has an affinity for heterochromatin conformation. Mol Cell Biol. 2007;27:1665–76.

[92] Meir Y-JJ, Weirauch MT, Yang H-S, Chung P-C, Yu RK, Wu SC-Y. Genome-wide target profiling of piggyBac and Tol2 in HEK 293: pros and cons for gene discovery and gene therapy. BMC Biotechnol. BioMed Central Ltd; 2011 Jan;11:28.

[93] Burnight ER, Staber JM, Korsakov P, Li X, Brett BT, Scheetz TE, et al. A hyperactive transposase promotes persistent gene transfer of a piggyBac DNA transposon. Mol Ther Nucleic Acids. 2012;1:e50.

[94] Galvan DL, Nakazawa Y, Kaja A, Kettlun C, Laurence JN, Rooney CM, et al. Genome-wide mapping of PiggyBac transposon integrations in primary human T cells. J Immunother. 2009;32:837–44.

[95] Garrison BS, Yant SR, Mikkelsen JG, Kay MA. Postintegrative gene silencing within the Sleeping Beauty transposition system. Mol Cell Biol. 2007;27:8824–33.

[96] Yusa K, Takeda J, Horie K. Enhancement of Sleeping Beauty transposition by CpG methylation: possible role of heterochromatin formation. Mol Cell Biol. 2004;24:4004–18.

[97] Sharma N, Hollensen AK, Bak RO, Staunstrup NH, Schrøder LD, Mikkelsen JG. The impact of cHS4 insulators on DNA transposon vector mobilization and silencing in retinal pigment epithelium cells. PLoS One. 2012;7:e48421.

[98] Rauschhuber C, Ehrhardt A. RNA interference is responsible for reduction of transgene expression after Sleeping Beauty transposase mediated somatic integration. PLoS One. 2012;7:e35389.

[99] Park CW, Park J, Kren BT, Steer CJ. Sleeping Beauty transposition in the mouse genome is associated with changes in DNA methylation at the site of insertion. Genomics. 2006;88:204–13.

[100] Zhu J, Park CW, Sjeklocha L, Kren BT, Steer CJ, Park CW, et al. High-level genomic integration, epigenetic changes, and expression of Sleeping Beauty transgene. Biochemistry. 2010;49:1507–21.

[101] Ley D, Harraghy N, Le Fourn V, Bire S, Girod P-A, Regamey A, et al. MAR elements and transposons for improved transgene integration and expression. PLoS One. 2013;8:e62784.

[102] Sjeklocha L, Chen Y, Daly MC, Steer CJ, Kren BT. β-Globin matrix attachment region improves stable genomic expression of the Sleeping Beauty transposon. J Cell Biochem. 2011;112:2361–75.

[103] Walisko O, Schorn A, Rolfs F, Devaraj A, Miskey C, Izsvák Z, et al. Transcriptional activities of the Sleeping Beauty transposon and shielding its genetic cargo with insulators. Mol Ther. 2008;16:359–69.

[104] Dalsgaard T, Moldt B, Sharma N, Wolf G, Schmitz A, Pedersen FS, et al. Shielding of Sleeping Beauty DNA transposon-delivered transgene cassettes by heterologous insulators in early embryonal cells. Mol Ther. 2009;17:121–30.

[105] Mossine VV, Waters JK, Hannink M, Mawhinney TP. piggyBac Transposon plus insulators overcome epigenetic silencing to provide for stable signaling pathway reporter cell lines. PLoS One. 2013;8:e85494.

[106] Sjeklocha LM, Park C-W, Wong PY-P, Roney MJ, Belcher JD, Kaufman DS, et al. Erythroid-specific expression of β-globin from Sleeping Beauty-transduced human hematopoietic progenitor cells. PLoS One. 2011;6:e29110.

[107] Bire S, Rouleux-bonnin F. Transgene Site-specific integration: problems and solutions. In: Renault S, Duchateau P, editors. Site-directed insertion of transgenes. Dordrecht, Netherlands: Springer; 2013:3–39.

[108] Ivics Z, Katzer A, Stüwe EE, Fiedler D, Knespel S, Izsvák Z. Targeted Sleeping Beauty transposition in human cells. Mol Ther. 2007;15:1137–44.

[109] Demattei M-V, Thomas X, Carnus E, Augé-Gouillou C, Renault S. Site-directed integration of transgenes: transposons revisited using DNA-binding-domain technologies. Genetica. 2010;138:531–40.

[110] Maragathavally KJ, Kaminski JM, Coates CJ. Chimeric Mos1 and piggyBac transposases result in site-directed integration. FASEB J. 2006;20:1880–2.

[111] Wu SC-Y, Meir Y-JJ, Coates CJ, Handler AM, Pelczar P, Moisyadi S, et al. piggyBac is a flexible and highly active transposon as compared to Sleeping Beauty, Tol2, and Mos1 in mammalian cells. Proc Natl Acad Sci. 2006;103:15008–13.

[112] Wilson MH, Kaminski JM, George AL. Functional zinc finger/Sleeping Beauty transposase chimeras exhibit attenuated overproduction inhibition. FEBS Lett. 2005;579:6205–9.

[113] Yant SR, Huang Y, Akache B, Kay MA. Site-directed transposon integration in human cells. Nucleic Acids Res. 2007;35:e50.

[114] Owens JB, Urschitz J, Stoytchev I, Dang NC, Stoytcheva Z, Belcaid M, et al. Chimeric piggyBac transposases for genomic targeting in human cells. Nucleic Acids Res. 2012;40:6978–91.

[115] Mandell JG, Barbas CF. Zinc finger tools: custom DNA-binding domains for transcription factors and nucleases. Nucleic Acids Res. 2006;34:W516–23.

[116] Kettlun C, Galvan DL, George AL, Kaja A, Wilson MH. Manipulating piggyBac transposon chromosomal integration site selection in human cells. Mol Ther. 2011;19:1636–44.

[117] Wang H, Mayhew D, Chen X, Johnston M, Mitra RD. "Calling cards" for DNA-binding proteins in mammalian cells. Genetics. 2012;190:941–9.

[118] Hickman AB, Dyda F. CRISPR-Cas immunity and mobile DNA: a new superfamily of DNA transposons encoding a Cas1 endonuclease. Mob DNA. 2014;5:23.

[119] Krupovic M, Makarova KS, Forterre P, Prangishvili D, Koonin EV. Casposons: a new superfamily of self-synthesizing DNA transposons at the origin of prokaryotic CRISPR-Cas immunity. BMC Biol. 2014;12:36.

[120] Xie F, Ye L, Chang JC, Beyer AI, Wang J, Muench MO, et al. Seamless gene correction of β-thalassemia mutations in patient-specific iPSCs using CRISPR/Cas9 and piggyBac. Genome Res. 2014;24:1526–33.

[121] Ye L, Wang J, Beyer AI, Teque F, Cradick TJ, Qi Z, et al. Seamless modification of wild-type induced pluripotent stem cells to the natural CCR5Δ32 mutation confers resistance to HIV infection. Proc Natl Acad Sci. 2014;111:9591–6.

[122] Toscano MG, Romero Z, Muñoz P, Cobo M, Benabdellah K, Martin F. Physiological and tissue-specific vectors for treatment of inherited diseases. Gene Ther. 2011;18:117–27.

[123] Liu L, Sanz S, Heggestad AD, Antharam V, Notterpek L, Fletcher BS. Endothelial targeting of the Sleeping Beauty transposon within lung. Mol Ther. 2004;10:97–105.

[124] Zhu J, Kren BT, Park CW, Bilgim R, Wong PY-P, Steer CJ. Erythroid-specific expression of beta-globin by the Sleeping Beauty transposon for Sickle cell disease. Biochemistry. 2007;46:6844–58.

[125] Hong I-S, Lee H-Y, Kim H-P. Novel therapeutic approaches for various cancer types using a modified Sleeping Beauty-based gene delivery system. PLoS One. 2014;9:e86324.

[126] Song JS, Kim HP, Rubin E. Development of a Sleeping Beauty-based telomerase gene delivery system for hepatocytes. Biosci Biotechnol Biochem. 2011;75:227–31.

[127] Chuah MK, Petrus I, De Bleser P, Le Guiner C, Gernoux G, Adjali O, et al. Liver-specific transcriptional modules identified by genome-wide *in silico* analysis enable efficient gene therapy in mice and non-human primates. Mol Ther. 2014;22:1605–13.

[128] Di Matteo M, Samara-Kuko E, Ward NJ, Waddington SN, Waddingon SN, McVey JH, et al. Hyperactive piggyBac transposons for sustained and robust liver-targeted gene therapy. Mol Ther. 2014;22:1614–24.

[129] Nakazawa Y, Huye LE, Salsman VS, Leen AM, Ahmed N, Rollins L, et al. PiggyBac-mediated cancer immunotherapy using EBV-specific cytotoxic T-cells expressing HER2-specific chimeric antigen receptor. Mol Ther. 2011;19:2133–43.

[130] Saito S, Nakazawa Y, Sueki A, Matsuda K, Tanaka M, Yanagisawa R, et al. Anti-leukemic potency of piggyBac-mediated CD19-specific T cells against refractory Philadelphia chromosome-positive acute lymphoblastic leukemia. Cytotherapy. 2014;16:1257–69.

[131] Tsukahara T, Iwase N, Kawakami K, Iwasaki M, Yamamoto C, Ohmine K, et al. The Tol2 transposon system mediates the genetic engineering of T-cells with CD19-specific chimeric antigen receptors for B-cell malignancies. Gene Ther. 2015;22:209–15.

[132] Kebriaei P, Huls H, Jena B, Munsell M, Jackson R, Lee DA, et al. Infusing CD19-directed T cells to augment disease control in patients undergoing autologous hematopoietic stem-cell transplantation for advanced B-lymphoid malignancies. Hum Gene Ther. 2012;23:444–50.

[133] Fjord-Larsen L, Kusk P, Emerich DF, Thanos C, Torp M, Bintz B, et al. Increased encapsulated cell biodelivery of nerve growth factor in the brain by transposon-mediated gene transfer. Gene Ther. 2012;19:1010–7.

[134] Bell JB, Podetz-Pedersen KM, Aronovich EL, Belur LR, McIvor RS, Hackett PB. Preferential delivery of the Sleeping Beauty transposon system to livers of mice by hydrodynamic injection. Nat Protoc. 2007;2:3153–65.

[135] Herweijer H, Wolff JA. Progress and prospects: naked DNA gene transfer and therapy. Gene Ther. 2003;10:453–8.

[136] Aronovich EL, Bell JB, Belur LR, Gunther R, Koniar B, Erickson DCC, et al. Prolonged expression of a lysosomal enzyme in mouse liver after Sleeping Beauty transposon-mediated gene delivery: implications for non-viral gene therapy of mucopolysaccharidoses. J Gene Med. 2007;9:403–15.

[137] Belur LR, Podetz-Pedersen KM, Sorenson BS, Hsu AH, Parker JB, Carlson CS, et al. Inhibition of angiogenesis and suppression of colorectal cancer metastatic to the liver using the Sleeping Beauty transposon system. Mol Cancer. 2011;10:14.

[138] Podetz-Pedersen KM, Bell JB, Steele TWJ, Wilber A, Shier WT, Belur LR, et al. Gene expression in lung and liver after intravenous infusion of polyethylenimine complexes of Sleeping Beauty transposons. Hum Gene Ther. 2010;21:210–20.

[139] Belur L, Frandsen JL, Dupuy AJ, Ingbar DH, Largaespada DA, Hackett PB, et al. Gene insertion and long-term expression in lung mediated by the Sleeping Beauty transposon system. Mol Ther. 2003;8:501–7.

[140] Chen S, Chen J, Huang P, Meng X-L, Clayton S, Shen J-S, et al. Myocardial regeneration in adriamycin cardiomyopathy by nuclear expression of GLP1 using ultrasound targeted microbubble destruction. Biochem Biophys Res Commun. 2015;458:823–9.

[141] Xue X, Huang X, Nodland SE, Mátés L, Ma LL, Izsvák Z, et al. Stable gene transfer and expression in cord blood-derived CD34+ hematopoietic stem and progenitor cells by a hyperactive Sleeping Beauty transposon system. Gene. 2010;114:1319–30.

[142] Ma K, Wang D-D, Lin Y, Wang J, Petrenko V, Mao C. Synergetic targeted delivery of Sleeping-Beauty transposon system to mesenchymal stem cells using LPD nanoparticles modified with a phage-displayed targeting peptide. Adv Funct Mater. 2013;23:1172–81.

[143] Tsukiyama T, Asano R, Kawaguchi T, Kim N, Yamada M, Minami N, et al. Simple and efficient method for generation of induced pluripotent stem cells using piggyBac transposition of doxycycline-inducible factors and an EOS reporter system. Genes Cells. 2011;16:815–25.

[144] Matsui H, Fujimoto N, Sasakawa N, Ohinata Y, Shima M, Yamanaka S, et al. Delivery of full-length factor VIII using a piggyBac transposon vector to correct a mouse model of hemophilia A. PLoS One. 2014;9:e104957.

[145] Liu L, Mah C, Fletcher BS. Sustained FVIII expression and phenotypic correction of hemophilia A in neonatal mice using an endothelial-targeted Sleeping Beauty transposon. Mol Ther. 2006;13:1006–15.

[146] Ohlfest JR, Frandsen JL, Fritz S, Lobitz PD, Perkinson SG, Clark KJ, et al. Phenotypic correction and long-term expression of factor VIII in hemophilic mice by immunotolerization and nonviral gene transfer using the Sleeping Beauty transposon system. Blood. 2005;105:2691–8.

[147] Yant SR, Meuse L, Chiu W, Ivics Z, Izsvak Z, Kay MA. Somatic integration and long-term transgene expression in normal and haemophilic mice using a DNA transposon system. Nat Genet. 2000;25:35–41.

[148] Chen ZJ, Kren BT, Wong PY-P, Low WC, Steer CJ. Sleeping Beauty-mediated down-regulation of huntingtin expression by RNA interference. Biochem Biophys Res Commun. 2005;329:646–52.

[149] Pan X-J, Ma Z-Z, Zhang Q-J, Fan L, Li Q-H. Sleeping Beauty transposon system is a reliable gene delivery tool for hereditary tyrosinaemia type 1 disease gene therapy: size of the foreign gene decides the timing of stable integration into the host chromosomes. J Int Med Res. 2012;40:1850–9.

[150] Montini E, Held PK, Noll M, Morcinek N, Al-Dhalimy M, Finegold M, et al. *In vivo* correction of murine tyrosinemia type I by DNA-mediated transposition. Mol Ther. 2002;6:759–69.

[151] Wilber A, Wangensteen KJ, Chen Y, Zhuo L, Frandsen JL, Bell JB, et al. Messenger RNA as a source of transposase for Sleeping Beauty transposon-mediated correction of hereditary tyrosinemia type I. Mol Ther. 2007;15:1280–7.

[152] Belcher JD, Vineyard JV, Bruzzone CM, Chen C, Beckman JD, Nguyen J, et al. Heme oxygenase-1 gene delivery by Sleeping Beauty inhibits vascular stasis in a murine model of sickle cell disease. J Mol Med (Berl). 2010;88:665–75.

[153] Sjeklocha LM, Wong PY-P, Belcher JD, Vercellotti GM, Steer CJ. β-Globin Sleeping Beauty transposon reduces red blood cell sickling in a patient-derived CD34(+)-based *in vitro* model. PLoS One. 2013;8:e80403.

[154] Aronovich EL, Bell JB, Khan SA, Belur LR, Gunther R, Koniar B, et al. Systemic correction of storage disease in MPS I NOD/SCID mice using the Sleeping Beauty transposon system. Mol Ther. 2009;17:1136–44.

[155] Aronovich EL, Hall BC, Bell JB, Mc Ivor RS, Hackett PB. Quantitative analysis of α-L-iduronidase expression in immunocompetent mice treated with the Sleeping Beauty transposon system. PLoS One. 2013;8:e78161.

[156] Yusa K, Rashid ST, Strick-Marchand H, Varela I, Liu P-Q, Paschon DE, et al. Targeted gene correction of α1-antitrypsin deficiency in induced pluripotent stem cells. Nature. 2011;478:391–4.

[157] Hyland KA, Olson ER, Clark KJ, Aronovich EL, Hackett PB, Blazar BR, et al. Sleeping Beauty-mediated correction of Fanconi anemia type C. J Gene Med. 2011;13:462–9.

[158] Wang X, Sarkar DP, Mani P, Steer CJ, Chen Y, Guha C, et al. Long-term reduction of jaundice in Gunn rats by nonviral liver-targeted delivery of Sleeping Beauty transposon. Hepatology. 2009;50:815–24.

[159] Ortiz-Urda S, Lin Q, Yant SR, Keene D, Kay MA, Khavari PA. Sustainable correction of junctional epidermolysis bullosa via transposon-mediated nonviral gene transfer. Gene Ther. 2003;10:1099–104.

[160] Bertino P, Urschitz J, Hoffmann FW, You BR, Rose AH, Park WH, et al. Vaccination with a piggyBac plasmid with transgene integration potential leads to sustained antigen expression and CD8(+) T cell responses. Vaccine. 2014;32:1670–7.

[161] Martinez-Fernandez A, Nelson TJ, Reyes S, Alekseev AE, Secreto F, Perez-Terzic C, et al. iPS cell-derived cardiogenicity is hindered by sustained integration of reprogramming transgenes. Circ Cardiovasc Genet. 2014;7:667–76.

[162] Tsukiyama T, Kato-Itoh M, Nakauchi H, Ohinata Y. A comprehensive system for generation and evaluation of induced pluripotent stem cells using piggyBac transposition. PLoS One. 2014;9:e92973.

[163] Yin J, Fan Y, Qin D, Xiaocui Bian X, Bi X. Generation and characterization of virus-free reprogrammed melanoma cells by the piggyBac transposon. J Cancer Res Clin Oncol. 2013;139:1591–9.

[164] Inada E, Saitoh I, Watanabe S, Aoki R, Miura H, Ohtsuka M, et al. PiggyBac transposon-mediated gene delivery efficiently generates stable transfectants derived from cultured primary human deciduous tooth dental pulp cells (HDDPCs) and HDDPC-derived iPS cells. Int J Oral Sci. 2015;7:144–54.

[165] He C-X, Shi D, Wu W-J, Ding Y-F, Feng D-M, Lu B, et al. Insulin expression in livers of diabetic mice mediated by hydrodynamics-based administration. World J Gastroenterol. 2004;10:567–72.

[166] Johnen S, Izsvák Z, Stöcker M, Harmening N, Salz AK, Walter P, et al. Sleeping Beauty transposon-mediated transfection of retinal and iris pigment epithelial cells. Invest Ophthalmol Vis Sci. 2012;53:4787–96.

[167] Ohlfest JR, Demorest ZL, Motooka Y, Vengco I, Oh S, Chen E, et al. Combinatorial antiangiogenic gene therapy by nonviral gene transfer using the Sleeping Beauty transposon causes tumor regression and improves survival in mice bearing intracranial human glioblastoma. Mol Ther. 2005;12:778–88.

[168] Kang Y, Zhang X-Y, Jiang W, Wu C-Q, Chen C-M, Gu J-R, et al. The piggyBac transposon is an integrating non-viral gene transfer vector that enhances the efficiency of GDEPT. Cell Biol Int. 2009;33:509–15.

[169] Bahrambeigi V, Ahmadi N, Moisyadi S, Urschitz J, Salehi R, Haghjooy Javanmard S. PhiC31/PiggyBac modified stromal stem cells: effect of interferon γ and/or tumor necrosis factor (TNF)-related apoptosis-inducing ligand (TRAIL) on murine melanoma. Mol Cancer. 2014;13:255.

[170] Huang X, Guo H, Kang J, Choi S, Zhou TC, Tammana S, et al. Sleeping Beauty transposon-mediated engineering of human primary T cells for therapy of $CD19^+$ lymphoid malignancies. Mol Ther. 2008;16:580–9.

[171] Singh H, Manuri PR, Olivares S, Dara N, Dawson MJ, Huls H, et al. Redirecting specificity of T-cell populations for CD19 using the Sleeping Beauty system. Cancer Res. 2008;68:2961–71.

[172] Maiti SN, Huls H, Singh H, Dawson M, Figliola M, Olivares S, et al. Sleeping beauty system to redirect T-cell specificity for human applications. J Immunother. 2013;36:112–23.

[173] Galvan DL, O'Neil RT, Foster AE, Huye L, Bear A, Rooney CM, et al. Anti-tumor effects after adoptive transfer of IL-12 transposon-modified murine splenocytes in the OT-I-melanoma mouse model. PLoS One. 2015;10:e0140744.

[174] Ramanayake S, Bilmon I, Bishop D, Dubosq M-C, Blyth E, Clancy L, et al. Low-cost generation of Good Manufacturing Practice-grade CD19-specific chimeric antigen

receptor-expressing T cells using piggyBac gene transfer and patient-derived materials. Cytotherapy. 2015;17:1251–67.

[175] Manuri PVR, Wilson MH, Maiti SN, Mi T, Singh H, Olivares S, et al. piggyBac transposon/transposase system to generate CD19-specific T cells for the treatment of B-lineage malignancies. Hum Gene Ther. 2010;21:427–37.

[176] Huye LE, Nakazawa Y, Patel MP, Yvon E, Sun J, Savoldo B, et al. Combining mTor inhibitors with rapamycin-resistant T cells: a two-pronged approach to tumor elimination. Mol Ther. 2011;19:2239–48.

[177] Huang G, Yu L, Cooper LJN, Hollomon M, Huls H, Kleinerman ES. Genetically modified T cells targeting interleukin-11 receptor α-chain kill human osteosarcoma cells and induce the regression of established osteosarcoma lung metastases. Cancer Res. 2012;72:271–81.

[178] Deniger DC, Yu J, Huls MH, Figliola MJ, Mi T, Maiti SN, et al. Sleeping Beauty transposition of chimeric antigen receptors targeting receptor tyrosine kinase-like orphan receptor-1 (ROR1) into diverse memory T-cell populations. PLoS One. 2015;10:e0128151.

[179] Jin Z, Maiti S, Huls H, Singh H, Olivares S, Mátés L, et al. The hyperactive Sleeping Beauty transposase SB100X improves the genetic modification of T cells to express a chimeric antigen receptor. Gene Ther. 2011;18:849–56.

[180] Xiao J, Meng X-M, Huang XR, Chung AC, Feng Y-L, Hui DS, et al. miR-29 inhibits bleomycin-induced pulmonary fibrosis in mice. Mol Ther. 2012;20:1251–60.

[181] Liu H, Liu L, Fletcher BS, Visner GA. Sleeping Beauty-based gene therapy with indoleamine 2,3-dioxygenase inhibits lung allograft fibrosis. FASEB J. 2006;20:E1694–703.

[182] Liu L, Liu H, Visner G, Fletcher BS. Sleeping Beauty-mediated eNOS gene therapy attenuates monocrotaline-induced pulmonary hypertension in rats. FASEB J. 2006;20:2594–1596.

[183] Liang M, Woodard LE, Liang A, Luo J, Wilson MH, Mitch WE, et al. Protective role of insulin-like growth factor-1 receptor in endothelial cells against unilateral ureteral obstruction-induced renal fibrosis. Am J Pathol. 2015;185:1234–50.

Chapter 4

Gene Correction Technology and Its Impact on Viral Research and Therapy

Guan-Huei Lee and Myo Myint Aung

Additional information is available at the end of the chapter

http://dx.doi.org/10.5772/62231

Abstract

Aims

1. To explain why gene correction technology is a useful tool for studying chronic/latent viral infections.

2. To explain how gene editing technology may facilitate or restrict virus replication and impact on future therapy.

3. To cite specific examples of how gene correction technology is being applied to target specific viruses, including HIV, hepatitis B virus (HBV), herpes simplex virus (HSV), and other viruses.

Methodology

We attempted to identify all scientific publications including basic science, translation research, and any clinical trials involving DNA correction technology [zinc finger endonuclease (ZFN), transcription activator-like endonucleases (TALENs), and CRISPR/Cas-based systems] and persistent viral infections [including but not limited to HIV, hepatitis B, C, and D viruses, herpes viruses, cytomegalovirus, Epstein–Barr virus (EBV), human papillomavirus (HPV), measles virus, and varicella-zoster virus] published on or before December 31, 2015. We conducted searches of MEDLINE, Cochrane Central Register, and EMBASE. The identified papers have been summarized and organized into relevant sections within the chapter.

Conclusion

Sequence-specific DNA endonucleases target and destroy DNA viruses, with early work describing the use of ZFNs, TALENs, or a third type of endonuclease, called a homing endonuclease (HE), to target HBV, HPV, and HSV-1 with varying degrees of success. The new CRISPR/Cas9 systems do not allow virologists to screen for host genes that affect the replication of pathogenic human viruses but to derive human cell lines that are genetically engineered to either facilitate or suppress viral replication. Scientists now widely use adeno-associated virus (AAV)-based vectors to directly target chronic viruses that infect discrete organs/tissues in the human body, such as HBV and HSV, although the safety of such delivery system is still a concern. The eventual goal is to eradicate or to disable the entire population of latent viral DNA genomes within the infected cells, resulting in permanent cure for these viral infections, which remains elusive until now.

Keywords: antiviral gene therapy, CRISPR/Cas, gene editing, sequence-specific DNA cleavage, TALENs, zinc finger

1. Introduction

Sequence-specific DNA endonucleases were first identified in the 1960s as enzymes that restrict the ability of DNA bacteriophage to grow in particular bacterial isolates or species. However, it was many years later that the DNA editing technology for selective modification of large viral or cellular DNA genomes was developed with the discovery of zinc finger endonuclease (ZFN) [1]. It was originally developed as an artificial restriction endonuclease to replace or complement restriction enzymes, and the engineered nucleases have become a versatile and indispensable tool in research as well as in biotechnology. However, each zinc finger is not necessarily highly specific for the 3-bp target that it is designed to bind and zinc fingers specific for all possible 3-bp target sequences have not yet been derived [2].

To address some of the deficiencies of ZFN, transcription activator-like endonucleases (TALENs) were developed based on a distinct modular DNA-binding motif. TALENs are made up of four ~34 amino acid domains derived from a DNA-binding protein found in the pathogenic plant bacterium *Xanthomonas*, which each recognizes a single DNA base pair with higher specificity [3, 4]. Although being more specific than ZFNs, TALENs require a rather laborious exercise in genetic engineering and are quite large, thus limiting the ability to express TALENs using viral vectors.

The newly developed CRISPR/Cas-based systems, or RNA-guided engineered nucleases (RGENs), unlike ZFNs and TALENs that use protein motifs for DNA sequence recognition, depend on RNA-DNA recognition. These are both highly specific and allow facile retargeting to new genomic loci [5, 6] and may be superseding the two older technologies. A key step forward in making the CRISPR/Cas9 system more user-friendly is by demonstrating that the crRNA and tracrRNA could be linked by an artificial loop sequence to generate a fully functional small guided RNA (sgRNA) [5, 6].

To date, the ever-increasing list of organisms with genomes that have been modified successfully using engineered nucleases includes mosquitoes, crickets, silkworms, pigs, cows, rabbits, and nonhuman primates, among others [7, 8]. The ability to genetically manipulate human pluripotent stem cells and somatic cells using engineered nucleases opens new opportunities to develop novel therapies for patients with various genetic and acquired diseases. Genetic defect correction is now possible in cultured cells from patients with a number of genetic diseases, including sickle cell disease [9], cystic fibrosis [10], Down syndrome [11], Duchenne muscular dystrophy [12], α1-antitrypsin deficiency [13], dystrophic epidermolysis bullosa [14], chronic granulomatous disease [15], and infectious diseases, particularly persistent viral infections that were among the most attractive targets in both ex vivo and in vivo studies, which will be discussed in the rest of the chapter. Despite all these promising advances, we should also be aware that these DNA editing technologies have been shown to cut at off-target sites with mutagenic consequences. Therefore, issues such as efficacy, specificity, and delivery are likely to drive selection of reagents for particular purposes. Human therapeutic applications of these technologies will ultimately depend on risk-versus-benefit analysis [8].

2. Main text

2.1. Why is DNA editing technology important in chronic viral disease research?

Chronic viral infection underlies a wide variety of medically important diseases that either follow directly from primary infection or may require months, years, or even decades to develop. Diseases caused by persistent virus infections include AIDS, AIDS-related complexes, chronic hepatitis, subacute sclerosing panencephalitis (chronic measles encephalitis), chronic papovavirus encephalitis (progressive multifocal leukoencephalopathy), several herpes virus–induced diseases, and some neoplasias (see **Table 1**). Pathogens with worldwide impact, such as HIV, hepatitis B virus (HBV), and a number of herpes viruses remain uncontrolled. The pathogenic mechanisms by which these viruses cause diseases include disorders of biochemical, cellular, immune, and physiologic processes. The chronic nature of these infections limits the number of antiviral options and increases the risk of developing drug-resistant strains in the host. Some recent studies suggest that chronic viral infection also contributes to certain cancers as well as to diabetes and atherosclerosis. Ongoing studies are rapidly advancing our understanding of many persistent infections. Viruses have evolved a wide variety of strategies by which they maintain long-term infection of populations, individuals, and tissue cultures.

Most of our current understanding of the host immunity and viral virulence are from studies on the progression of acute infection. When a virus enters the host, there is an initial nonequilibrium phase of acute infection, in which viral and immune strategies compete for dominance. A transition point will be reached in a survivor, at which the infection either is cleared or becomes chronic. This transition point may be reached very early in infection for viruses that can establish a latent infection, in which case the infection is permanent regardless of the course of acute infection. If recovery occurs, the immune system must reset by clearing the antigen and reestablishing immune homeostasis. If the balance shifts toward chronic infection, a new set of viral and host strategies interact to define a metastable equilibrium in which viral replication is held in check, but the virus is not cleared [16].

Whereas acute viral infection represents a nonequilibrium process, chronic viral infection is a process in dynamic and metastable equilibrium. During acute infection, both the host and the virus change continuously until infection is resolved, kills the host, or becomes chronic. Certain genes in a virus or in the immune system function during acute but not chronic infection. The failure of these immune system genes to function effectively or the overly effective evasion of immunity by the virus may eventually lead to multiorgan failure and death. In contrast, during chronic infection, viral and host genes balance each other [16]. The mechanisms of viral persistence despite the impressive immune armamentarium of the host, without causing overt disease, remain unclear, although the events leading to the establishment of chronicity were described.

Two events are fundamental to the establishment of chronic viral infection. First, the virus must evade sterilizing immunity (the complete elimination of a virus). Second, the immune system must adjust to the continuous presence of viral antigen-driven inflammatory responses to limit viral replication to an acceptable level without untoward damage to permanently infected tissues. If the immune system cannot eliminate the virus, unrestrained immune attack on virus antigen-bearing cells causes tissue injury. Thus, down-regulation of inflammation during chronic viral infections can result in decreased tissue damage, at least for noncytopathic viruses. Immunopathology can be severe in human chronic infections caused by HBV and hepatitis C virus (HCV) [46], which may remain noncytopathic for many decades initially. It is important to realize that viruses which rely on a living but chronically infected host for their own survival must carefully avoid mechanisms that overwhelm immunity and kill their hosts.

Viruses have evolved highly effective strategies for establishing chronic infection despite the presence of an active host antiviral immune response. There are three general strategies for chronic viral infection: continuous replication, latency and reactivation, and invasion of the genome followed by vertical spread from generation to generation. Individual viruses usually rely mostly on one strategy, but viruses can use more than one mechanism. For example, HIV effectively uses both continuous replication and establishment of latency [47–49], a dangerous combination. The differences between these strategies have profound implications for designing new ways to prevent or control harmful chronic viral infections.

The mechanisms of the viral persistence have not been completely understood, and some common factors are known [50]. Immune modulation is one of them. Many of these viruses managed to avoid the specific and nonspecific immune defenses in several different ways. These include (1) limitation of recognition molecules on infected cells; (2) altered lymphocyte and macrophage functions, including the modified production of cytokines and general immunosuppression [e.g., HIV-1 and -2, Epstein–Barr virus (EBV), and HBV]; (3) infection in immunologically privileged anatomic sites [e.g., herpes simplex virus (HSV) and VZV in central nervous system]; (4) compromised nonspecific defenses (e.g., suppress interferon production); and (5) immune tolerance (e.g., HBV).

Another common mechanism is the modulation of viral gene expression. Such examples include (1) down-regulation of some viral genes by viral or cellular regulatory gene products [e.g., HIV and human papillomaviruses (HPVs)], (2) specific latency-associated proteins (e.g., EBNA-1), and (3) synthesis of latency-associated transcripts (LATs; e.g., HSV-1 and -2) as well as viral variants (e.g., HIV and measles) [50].

Developing a successful cure for persistent or chronic viral infections remains a major challenge due to their ability to evade/suppress the immune system and their ability to incorporate viral sequences into the host genome and long inactive latent phases, which makes targeting of active biological activities nearly impossible. However, recent concerted effort and improvement of drug design, leading to multiple new drugs that successful cure HCV, demonstrate that the complete clearance of chronic viral infections is an attainable goal. A particularly tantalizing application of programmable nucleases is the potential to directly correct genetic mutations in affected tissues and cells to treat diseases that are refractory to traditional therapies. A number of approaches targeting specific disease-causing viral infections will be discussed in the following sections.

			Disease during chronic infection		
Virus, primary nucleic acid, estimated percent of humans infected	**Major site of persistence (organ or cell)**	**Acute infection examples**	**Within normal hosts**	**Within immuno compromised hosts**	**References**
Hepatitis B virus (HBV), DNA, 350 million, ~5%	Hepatocytes	Hepatitis	Cirrhosis, hepatocellular carcinoma	Same diseases	McGovern [17]; Rehermann and Nascimbeni [18]
Hepatitis C virus (HCV), RNA, 170 million, ~2.5%	Hepatocytes	Hepatitis	Cirrhosis, hepatocellular carcinoma	Same diseases	Rehermann and Nascimbeni [18]; Lemon et al. [19]
Human immunodeciency virus (HIV-1 and HIV-2), RNA, 33 million, ~0.5%	CD4+ T cells, monocyte/ macrophages	Acute febrile illness	AIDS	AIDS	UNAIDS [20]; Kuritzkes and Walter [21]
Hepatitis D virus (HDV), RNA, 15 million, ~0.2%	Hepatocytes	Unknown	Exacerbation of chronic HBV infection	Unknown	Taylor et al. [22]
Human T cell leukemia virus type 1 (HTLV1), RNA, 10–20 million, ~0.2%	T cells	Unknown	Adult T cell leukemia (2–6% of carriers), tropical spastic para-paresis, myelopathy, uveitis, dermatitis	Unknown	Matsuoka and Jeang [23]; Lairmore and Franchini [24]
Xenotropic murine leukemia virus-related	Prostate	Unknown	Prostate cancer?	Unknown	Urisman et al. [25]; Dong et al. [26]

Virus, primary nucleic acid, estimated percent of humans infected	Major site of persistence (organ or cell)	Acute infection examples	Disease during chronic infection		References
			Within normal hosts	Within immuno compromised hosts	
virus (XMLV), RNA, unknown					
Rubella virus, German measles, RNA, rare	CNS	Rubella, arthritis	Progressive rubella panencephalitis	Unknown	Hobman and Chantler [27]
Parvovirus B19, DNA, rare	Bone marrow erythroid progenitors	Fifth disease, arthritis	Aplastic crisis in hemolytic anemia, hydrops fetalis, chronic bone marrow de ciency	Red cell aplasia	Berns and Parrish [28]; Norja et al. [29]
Human papillomavirus, DNA, <5%	Epithelial skin cells	Unknown	Papilloma, cervical and other mucosal carcinomas	Increased severity and incidence of same diseases	Leggatt and Frazer [30]; Howley and Lowy [31]
Measles virus, RNA, rare	Neurons and supporting cells in CNS	Measles	Subacute sclerosing panencephalitis, measles inclusion body encephalitis	Unknown	Griffin [32]
Coxsackie, RNA, rare	Myocardial cells	Hand foot and mouth disease, herpangina	Myocarditis	Unknown	Chapman and Kim[33]; Whitton and Feuer [34]
Human herpes virus 6 (HHV-6), DNA, >90%	Lymphocytes?	Roseola	Unknown	Meningoencephalitis, secondary infections, immunomodulatory?	Straus[35]; Yamanishi et al. [36]
Varicella zoster virus (VZV), DNA, >90%	Sensory ganglia neurons and/or satellite cells, lymphocytes	Chicken pox	Herpes zoster	Disseminated disease, hepatitis, pneumonitis	Zerboni and Arvin [37]; Straus [38]
Cytomegalovirus (CMV), DNA, 80%–90%	Myelomonocytic cells	Mononucleosis	Rare	Disseminated disease, vasculitis, pneumonitis, retinitis, hepatitis,	Mocarski et al. [39]

Virus, primary nucleic acid, estimated percent of humans infected	Major site of persistence (organ or cell)	Acute infection examples	Disease during chronic infection		References
			Within normal hosts	Within immuno compromised hosts	
				gastroenteritis, meningoencephalitis	
Epstein–Barr virus (EBV), DNA, 80–90%	Pharyngeal epithelial cells, B cells	Mononucleosis	Burkitt's lymphoma, nasopharyngeal carcinoma, non-Hodgkin's lymphoma	CNS lymphomas, oral hairy leukoplakia, lymphoproliferative disease	Rickinson and Kieff [40]; Straus [38]; Kieff and Rickinson [41]
Polyomavirus BK, DNA, 72–98%	Kidney	Unknown	Unknown	Hemorrhagic cystitis (post bone marrow transplantation), nephropathy (post kidney transplantation)	Zur [42]
Polyomavirus JC, DNA, 72–98%	Kidney, CNS	Unknown	Unknown	Progressive multifocal leukoencephalopathy	Zur [42]
Herpes simplex type 1 (HSV-1), DNA, 50–70%	Sensory ganglia neurons	Pharyngitis, encephalitis, keratitis	Cold sores, encephalitis, keratitis	Increased severity of same diseases, pneumonitis, hepatitis	Straus [38]
Adenovirus, DNA, up to 80%	Adenoids, tonsils, lymphocytes	Upper respiratory infection, gastroenteritis	Unknown	Enteritis, hemorrhagic cystitis, pneumonitis, hepatitis, others	Garnett et al. [43]; Wold and Horwitz [44]
Herpes simplex type 2 (HSV-2), DNA, 20–50%	Sensory ganglia neurons	Genital herpes	Genital herpes, encephalitis	Increased severity of same diseases	Straus [38]
Kaposi's sarcoma herpes virus (KSHV) or human herpes virus 8, DNA, 2–60%	Endothelial cells, B cells	Unknown	Castleman's disease, Kaposi's sarcoma	Kaposi's sarcoma, primary effusion lymphoma	Ganem [45]

Table 1. List of known human pathogenic chronic viral infections and their respective prevalence and clinical disease [16].

2.2. Defining the cellular factors that facilitate or restrict virus replication

Using sequence-specific DNA endonucleases to target and destroy DNA viruses had been attempted almost as soon as the technology was invented [51], with earlier work describing the use of ZFNs, TALENs, or an earlier HE, to target HBV, HPV, and HSV-1.

The newer DNA editing systems not only impact virology by allowing researchers to screen for human genes that affect the replication of pathogenic human viruses and are extremely useful in generating human cell lines that lack gene products that may facilitate or restrict specific virus replication. Adeno-associated virus (AAV)-based vectors offer the possibility of directly targeting DNA viruses that infect specific organs or tissues in the human body (e.g., HSV and HBV) to target and destroy the entire population of viral DNA genomes. Safety concerns, such as off-target genomic mutations, of such delivery system remains [52].

Viruses have very compact genomes and therefore rely on the host cell for many activities required to support their replication cycle. Ironically, viruses as pathogens put selective pressure on their host organisms, thus selecting for cellular restriction factors that can limit its own level of viral replication. One of the best studied viruses in this context is HIV, which requires a wide range of human cofactors for replication in $CD4^{+}$ T cells, several of which are lacking in other mammalian species, such as in mice. HIV is targeted by a wide range of human restriction factors and dedicates a substantial portion of its coding capacity to the neutralization of these restriction factors. For example, the Vif protein blocks the activity of the host APOBEC3 family of restriction factors, which otherwise interfere with the production of the HIV-1 provirus, whereas the HIV-1 Vpu protein neutralizes the cellular restriction factor tetherin, which blocks the release of progeny virions [53].

The identity of cellular factors that either facilitate or restrict viral replication is central to the search for novel targets for antiviral drug development. Researchers have been using RNA interference (RNAi) libraries to knock down the expression of human genes systematically to identify factors required for virus replication. Although this approach has led to some interesting insights, it is also clear that RNAi screens for viral cofactors in different laboratories have often led to very different lists of cellular proteins with this potential activity. This may be due to the use of different cell systems, different assays for viral replication, different RNAi reagents, and sometimes incomplete knockdown of target gene expression. As a result, most of the viral cofactors and restriction factors identified so far have required confirmation using biochemical approaches (e.g., by identification of cellular factors that specifically bind to a viral protein) or genetic approaches (e.g., by complementation of a human and/or animal cell line that lacks a particular cofactor or restriction factor) [52, 53].

RGEN systems appear highly suitable for use in screens for viral cofactors or restriction factors; indeed, several screens for cellular factors involved in cell transformation have been published [54–56]. This could now be extended to analysis of the replicative potential of viruses in a 96- or 384-well plate format, possibly in the form of comprehensive CRISPR/Cas9-generated libraries of cellular clones available for specific human cell lines. Each of these lines will lack a functioning gene that is dispensable for cell viability in a tissue culture setting. These clones should allow the reproducible and almost complete identification of cellular factors that either

http://dx.doi.org/10.5772/62231

help or hinder virus replication in that human cell line in vitro. Factors required for human cell viability would be missed, but it seems likely that almost all host innate immune factors involved in restricting virus replication would not be essential. Essential factors required for host cell viability must also be essential for virus replication and clearly would not provide potential targets for antiviral drug development. A summary of studies using DNA editing technology to define the host factors that facilitate or restrict virus replication is listed in **Table 2**.

Defining the host factors that facilitate or restrict virus replication

Virus	Human disease	Type	Host factors studied	Cell/ animal/ human study	Delivery method	Reference
Cytomegalovirus (CMV)	Disseminated disease, vasculitis, pneumonitis, retinitis, hepatitis, gastroenteritis, meningo encephalitis	TALEN	Glucocorticoid receptor (GR) gene	CD8 + T cells	Electroporation	Menger et al. [57]
Hepatitis C virus (HCV)	Cirrhosis, hepatocellular carcinoma	TALEN	DGAT1	Huh7.5 cells	Transfection	Sung et al. [58]
		CRISPR	CLDN1, OCLN and CD81	Huh7.5 cells		Ren et al. [59]
		CRISPR	ISG15	Huh7 cells	Transfection	Domingues et al. [60]
Herpes simplex type 1 (HSV-1)	Core sores, encephalitis, pneumonitis, hepatitis	CRISPR	IFI16	U2OS cells	Transfection	Johnson et al. [61]
		CRISPR	LULL1	HeLa cells	Transfection	Turner et al. [62]
Human immuno deficiency virus	AIDS	ZFN	CCR5	CD4 + T cells, human	AAV	Perez et al. [63], Tebas et al. [64]

Defining the host factors that facilitate or restrict virus replication						
Virus	**Human disease**	**Type**	**Host factors studied**	**Cell/ animal/ human study**	**Delivery method**	**Reference**
(HIV-1)						
		CRISPR	CCR5	CD4 + T cells	Ad5F35 adenoviral vector	Li et al. [65]
Kaposi's sarcoma herpes virus (KSHV) or human herpes virus 8	Kaposi's sarcoma, primary effusion lymphoma	CRISPR	ORF45, RSK	SLK-iBAC cells	Lentiviral delivery method	Avey et al. [66]
Human papillo mavirus	Papilloma, cervical and other mucosal carcinoma	TALEN	E6, E7 ORFs	SiHa or HeLa cells, Mice	Transfection	Hu et al. [67]
		CRISPR	E6, E7 ORFs	HeLa cells	Transfection	Kennedy et al. [68]

Table 2. Summary of studies using DNA editing technology to define the host factors that facilitate or restrict virus replication.

2.3. Using engineered nucleases for gene therapy applications in specific viral pathogens

2.3.1. Human immunodeficiency virus (HIV)

In the last two decades, the availability of highly active antiretroviral therapy (HAART) capable of reducing HIV replication to undetectable level and prolonged survival has greatly improved the prognosis of infected patients. However, HIV-1 persists as a latent infection in a small number of resting $CD4^+$ memory T cells [47]. In these long-lived cells, intact integrated HIV-1 proviruses persist in a transcriptionally silent state that is refractory to both drugs and host immune responses. However, these memory T cells can be reactivated by an appropriate recall antigen, resulting in the induction of a productive viral replication cycle [47, 49]. If this occurs after drug treatment has been stopped, HIV-1 will rapidly spread through the available $CD4^+$ T cells and rekindle the same level of virus replication that was seen before antiviral drug treatment. Other disadvantages of such long-term therapy include limitations including high cost, patient compliance and side effects of long-term therapy, as well as emergence of drug resistance [84]. Therefore, there is a need to develop a more effective "cure" for HIV infection.

The approaches to purge the pool of latently infected cells have focused on two strategies. Some have attempted to activate latent HIV-1 proviruses using drugs, including histone deacetylase inhibitors and protein kinase C agonists [85]. However, so far, this strategy has not been proven able to activate HIV-1 in a high percentage of latently infected cells.

An alternative strategy would be to directly target and destroy latent proviruses using HIV-1-specific CRISPR/Cas combinations. Early attempts have shown that this is feasible and that latent proviruses can be excised from the host cell genome, and then destroyed, by cleavage in the HIV-1 long terminal repeat regions [86, 87]. In principle, the HIV-1 provirus is a perfect target for CRISPR/Cas, as there is only a single proviral copy in the infected cell, and in the presence of antiviral drugs, no spread of the virus is possible. The problem, however, is that latently HIV-1-infected T cells are scattered throughout the body and infecting all of these seems currently to be an insurmountable problem, especially as T cells are poor targets for AAV infection. This contrasts with HBV, HSV, and HPV, all of which are tightly localized in known tissues in the body that can be readily targeted by AAV [52]. Therefore, in the absence of novel vector delivery systems that can target latently HIV-1-infected cells throughout the body, HIV-1 is likely to remain a technically challenging target for elimination by CRISPR/Cas in vivo.

CCR5, which encodes a coreceptor for HIV entry [88, 89], has been a popular target for developing a new generation of HIV therapy for several reasons. First, its disruption seemed likely to increase the survival of CD4T cells; persons homozygous for a naturally occurring 32-bp deletion (delta32/delta32) in CCR5 are known to be resistant to HIV infection [90]. CD4 T cells from such persons are highly resistant to infection in vitro [91]. Persons who are heterozygous for CCR5 delta32 and HIV infection have slower progression to full-blown AIDS [92, 93]. The effectiveness of blocking or inhibiting CCR5 with the use of small interfering (siRNA) and other small-molecule inhibitors has been shown in humans [94]. Finally, there is the remarkable success story of HIV eradication in the so-called "Berlin Patient". This HIV-positive patient with lymphoma, who had been transplanted with bone marrow from a CCR5-Δ32 homozygous donor, became cured with no measurable virus (undetectable HIV RNA and proviral DNA in the blood, bone marrow, and rectal mucosa) even 5 years after transplantation, showing the potential benefits of CCR5 disruption [95, 96]. Although the mechanism responsible for the apparent cure associated with this procedure remains to be established, acquired CCR5 deficiency is one possibility [97]. Due to the low frequency of CCR5-Δ32 homozygotes in the general population and the difficulties of identifying suitable donors, alternative methods to artificially disrupt CCR5 are being sought [65]; in particular, gene editing methods have attracted a lot of attention as a potential therapy for HIV, as they allow permanent disruption of the selected gene(s).

This approach had recently led to a phase I clinical study published in the *New England Journal of Medicine* [64]. The researchers enrolled 12 patients in an open-label, nonrandomized, uncontrolled study of a single dose of ZFN-modified autologous CD4 T cells. Six of these 12 patients underwent an interruption in antiretroviral treatment 4 weeks after the infusion of 10 billion autologous CD4 T cells. Between 11% and 28% of these CD4 T cells were genetically modified by ZFN. The primary outcome was safety in terms of treatment-related adverse events. Secondary outcomes included measures of immune reconstitution and HIV resistance.

The only serious adverse event associated with infusion of the ZFN-modified CD4 T cells was attributed to a transfusion reaction. The median concentration of CCR5-modified CD4 T cells at 1 week was 250 cells/mm^3. This represented approximately 8.8% of circulating peripheral blood mononuclear cells and 13.9% of circulating CD4 T cells. These modified T cells had an estimated mean half-life of 48 weeks. During treatment interruption and the resultant viremia, the decline in circulating CCR5-modified cells (-1.81 cells/day) was significantly less than the decline in unmodified cells (-7.25 cells/day; $p = 0.02$; see **Figure 1**) [64]. HIV RNA became undetectable in one patient. The blood level of HIV RNA decreased in the majority of the study subjects. This provides one of the first evidence that CCR5-modified autologous CD4 T-cell infusions are safe and opens the door for applying DNA editing technologies for treating human chronic viral diseases, although many obstacles remain.

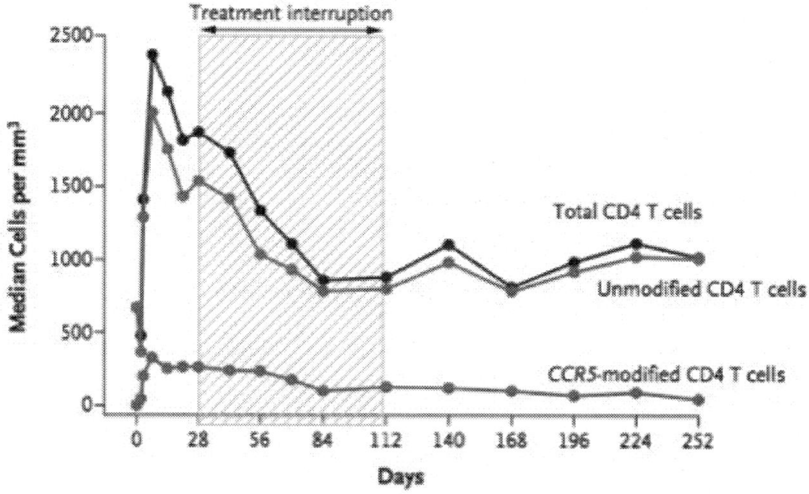

Figure 1. CCR5-modified CD4 T cells during treatment interruption [64].

2.3.2. *Hepatitis B virus (HBV)*

HBV remains a major public health problem, with more than 300 million people chronically infected worldwide [98]. These individuals have an approximately 25% risk of dying from the consequences of HBV infection, including hepatocellular carcinoma (HCC) and cirrhosis, and approximately 800,000 individuals are thought to die each year due to HBV. An effective vaccine for HBV is available, but this is not helpful in individuals with preexisting infection and is not fully effective at preventing vertical transmission, compounded by the problem that it is not given to all children in resource-limited regions timely. HBV polymerase can be effectively inhibited by nucleoside-based antiviral agents (lamivudine, adefovir, telbivudine, entecavir, tenofovir), but this does not cure this infection due to the extraordinary stability of the viral episomal cccDNA intermediate [99], which continue to persist and produce new viral particles, as soon as the antiviral agents were stopped.

Current research in HBV therapy focuses on finding new targets on the viral life cycle (see **Figure 2**) and trying to overcome immune tolerance through immunotherapy, with limited success. It is clear that a complete "cure" cannot be achieved unless the new therapy is able to remove or destabilize the HBV cccDNA, and genome editing is one of the most promising approaches. Some of the challenges that need to be overcome include toxicity, development of viral resistance, specificity, ensuring therapeutic effect of sufficient duration, and hepatocyte-targeted delivery. Significant progress has been made to overcome these obstacles.

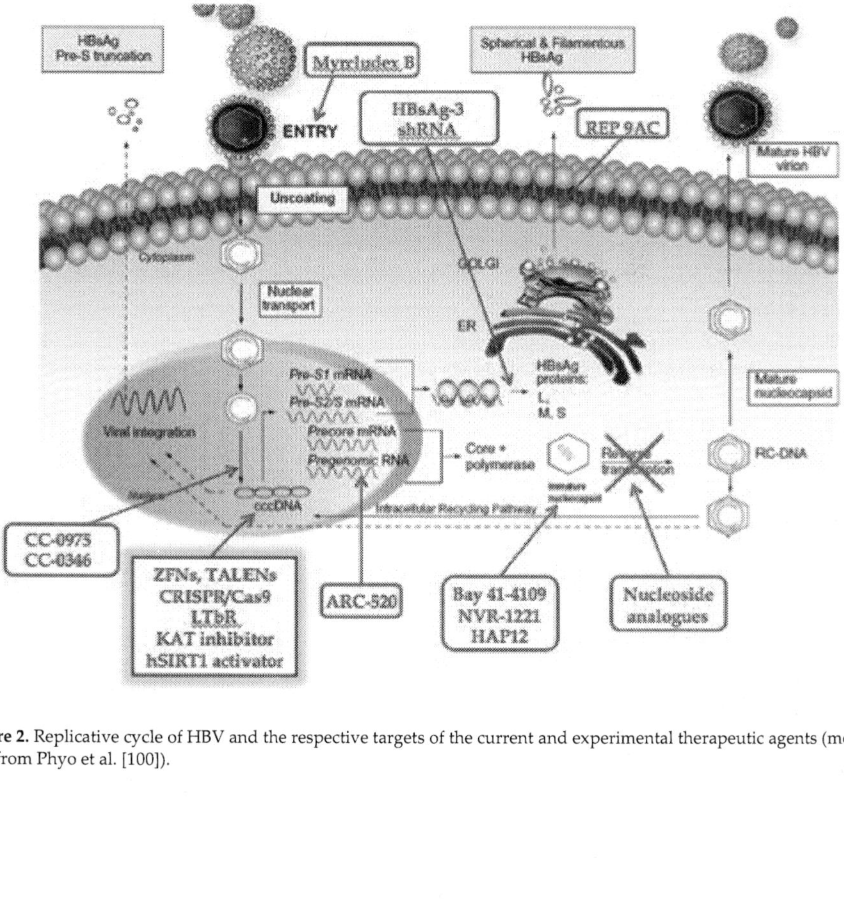

Figure 2. Replicative cycle of HBV and the respective targets of the current and experimental therapeutic agents (modified from Phyo et al. [100]).

Zimmerman et al. [101] were the first to advance a gene editing approach to countering HBV replication. The researchers used duck HBV (DHBV) as a model and designed six different zinc finger proteins (ZFPs) to target the DHBV enhancer sequences, which control the transcription of the core and surface sequences. Marked reduction in viral pgRNA and total viral RNA was observed. ZFPs significantly reduced viral core and surface protein production with no obvious cytotoxicity. As the ZFPs did not cause target DNA mutation or durable epigenetic changes, this inhibition was not lasting. Later, Cradick et al. [71] also demonstrated the effectiveness of using ZFNs to specifically cleave HBV episomal DNA. The team engineered nine pairs of HBV-specific ZFNs and cotransfected each of these ZFN pair plus an HBV genome target plasmid into a hepatoma cell line. Targeted cleavage of viral DNA was demonstrated. The cleaved fragments were misrepaired in a manner that could potentially inactivate HBV. Further, cotransfection with the ZFN pair 6 decreased HBV pregenomic viral RNA levels by almost 30%. However, it should be noted that the study did not clearly demonstrate that the HBV cccDNA was modified by the ZFNs.

The X gene is thought to play a major role in the development of HCC [102]. Zhao et al. [103] designed ZFPs to inhibit the expression of integrated sequences of the X gene. An artificial transcription factor (ATF) was synthesized to target a sequence in the enhancer I region, which is upstream of the X promoter. The ATF comprised a DNA-binding domain of a ZFP that was linked to a KRAB repressor domain. X repression was demonstrated on a luciferase reporter assay. Another study by Weber et al. [72] aimed to prevent viral reactivation by targeting three HBV protein-coding sequences with ZFNs in HepAD38 cells, a tet-regulated cell line. AAV vectors containing sequences encoding the ZFNs were used for gene delivery. Site-specific mutagenesis with low cytotoxicity was confirmed for two of three engineered ZFNs. Inhibition of HBV replication and virion production over a period of 14 days could be achieved after a single treatment with ZFN targeting the viral polymerase gene.

Bloom et al. [73] first applied TALENs to disrupt hepatitis B replication in a cell line and a mouse model. HBV-specific TALENs were generated, targeting conserved sequences in the surface, core, and pol open reading frames (ORFs). The TALEN targeting surface and core ORFs were the most effective. The viral cccDNA were isolated using Hirt's extraction and plasmid-safe DNase. T7 endonuclease I assay was used to verify targeted mutation in cultured cells. The TALEN targeting surface ORF reduced HBsAg by more than 90% and circulating viral particle equivalents were diminished by approximately 70% by the S and C TALENs. T7E1 assays and deep sequencing confirmed the targeted disruption. Chen et al. [74] confirmed the successful targeting and inactivation of HBV genomic sequences by TALENs. The researchers also showed significant knockdown in markers of viral replication. Interestingly, when used in combination with interferon-α, synergistic antiviral effects were observed. Although promising, a limitation of using mice to simulate HBV replication in vivo is that these animals do not produce HBV cccDNA.

Some key studies employing CRISPR/Cas9 systems recently demonstrated the utility of RGEN cleavage of HBV DNA [75–78]. Lin et al. [75] designed eight HBV-targeting sgRNAs to target different conserved regions of the HBV genome. A significant decrease in the production of viral proteins was observed. Cotransfection with more than one sgRNA-encoding sequence augmented antiviral efficacy. This effect was corroborated by an increase in indels at the targeted sites. However, the efficacy against HBV cccDNA was not evaluated. Seeger and Sohn [76] investigated the targeted disruption of HBV cccDNA and confirmed the efficient cleavage of viral sequences with all five of their sgRNA constructs. Approximately eightfold reduction of HBcAg expression was achieved in HBV-infected HepG2-NTCP cells. The cells were transduced using recombinant lentiviral vectors. Targeted deletions from single nucleotide up to 2.3 kb was produced. This demonstrated the potential of CRISPR/Cas to target and excise host-integrated HBV genomes. Another similar study by Kennedy et al. [77] also showed suppression of HBV replication by lentiviral vector-delivered Cas9 and sgRNA sequences. Dong et al. confirmed the efficacy of sgRNA-Cas9 against HBV. In addition, they demonstrated the disruption of artificial cccDNA in a murine hydrodynamic model [78] that was based on the use of engineered recombinant cccDNA precursor plasmid (rcccDNA) [104].

These studies suggest that HBV-specific Cas9/sgRNA combinations can block HBV replication and eliminate the cccDNA pool if they can be effectively delivered to hepatocytes. At the moment, AAV may be the best carrier for this task, as several AAV serotypes are naturally hepatotropic and even more highly hepatotropic AAV vectors have recently been isolated by "shuffling" AAV sequences in vivo [105]. The next step is to examine whether AAV-delivered Cas9/sgRNA combinations can effectively cure HBV in the humanized, immunodeficient mouse liver model system.

2.3.3. Herpes simplex virus (HSV)

HSV-1 infects approximately 70% of the U.S. population and about one third of affected individuals suffer from recurrent, primarily oral, cold sores. HSV-1 most commonly initially infects the oral mucosal epithelium, leading to a local productive infection, and then undergoes retrograde transport to the trigeminal ganglia, where it establishes a latent infection in a small number of sensory neurons, which persists after the initial, productive infection is cleared by the host immune response [106]. During latency, the HSV-1 DNA genome is maintained as a nuclear episome, with 1 to ~50 copies per latently infected neuron. At this point, the only region of the genome that is actively transcribed encodes the LAT, which is processed to give rise to a single long noncoding RNA of 2.1 kb, as well as eight virally encoded miRNAs, which together are thought to regulate exit from latency [107]. Because no viral proteins are made, there is no immune recognition of latently infected cells. Occasionally, one or more latently infected neuron is activated to produce infectious virions that migrate down the axons of the reactivating neuron to the original site of infection, where they reestablish a transient productive infection that can lead to the formation of cold sores. Although often no more than an irritation, HSV-1 infections can also lead to serious morbidity and HSV-1 keratitis represents the most common form of infectious blindness in the West [108]. Infection of the central nervous system may also cause fatal encephalitis. A closely related virus, HSV-2, which is found in approximately one fifth of the U.S. population, has a similar replication cycle but generally is sexually transmitted and infects the genital mucosa. Latency is established in sensory neurons of the sacral ganglia and reactivation can lead to genital ulcers. Again, serious morbidity is rare but does occur in some individuals and neonatal HSV-2 infections acquired during vaginal delivery can be fatal [106].

Although there are several drugs that can treat productive HSV-1 or -2 infections, generally by targeting viral DNA synthesis, latent HSV genomes are entirely refractory to current treatment regimens and it remains impossible to cure these infections. What is clearly needed is an approach that directly targets HSV-1 or HSV-2 episomal DNA for cleavage and elimination from latently infected neurons. AAV-delivered HSV-specific engineered endonucleases appear ideal for this purpose. Aubert et al. [79] introduced DNA double-stranded breaks in an HSV latency model using the engineered HE HSV1m5, which targets a sequence in the HSV-1 gene UL19, encoding the viral protein VP5. Coexpression of the 3′-exonuclease Trex2 with HEs increased the mutagenesis frequencies by up to sixfold. Following HSV1m5/Trex2 delivery with AAV vectors, the target site within latent HSV genomes was mutated. There was no detectable cell toxicity. The viral production by latently infected cells after reactivation was significantly decreased. Prior HSV1m5/Trex2 treatment followed by exposure to histone deacetylase inhibitors increased mutagenesis frequencies of latent HSV genomes by another twofold to fivefold. This indicates that chromatin modification may be a useful adjunct to gene-targeting methods. Using CRISPR/Cas9-mediated genome engineering to create single- and double-knockout (KO) cell lines, Turner et al. [62] discovered that the Torsin Activator LULL1 is required for efficient growth of HSV-1, whereas Johnson et al. [61] reported that the interferon-γ-inducible factor 16 (IFI16) restricts HSV-1 replication by accumulating the HSV-1 genome and repressing HSV-1 gene expression, and modulates histone modifications. Given the tight localization of HSV-1 and -2 to the trigeminal and sacral ganglia, respectively, low level of viral DNA genomes present in these cells, and the ability to efficiently transduce sensory neurons with AAV8-based vectors, this seems like an ideal viral candidate for cure using RGENs.

2.3.4. Human papillomavirus (HPV)

HPV infection, although normally innocuous, can also give rise to warts on the skin or genitalia [31]. Most HPV variants replicate as episomes in the basal epithelial layer of the skin, where the virus expresses exclusively nonstructural proteins. When the infected precursor epithelial cell migrates toward the surface of the epidermis and undergoes differentiation into a keratinocyte, the productive HPV replication cycle is activated leading to the release of infectious HPV virions (**Figure 3**) [31].

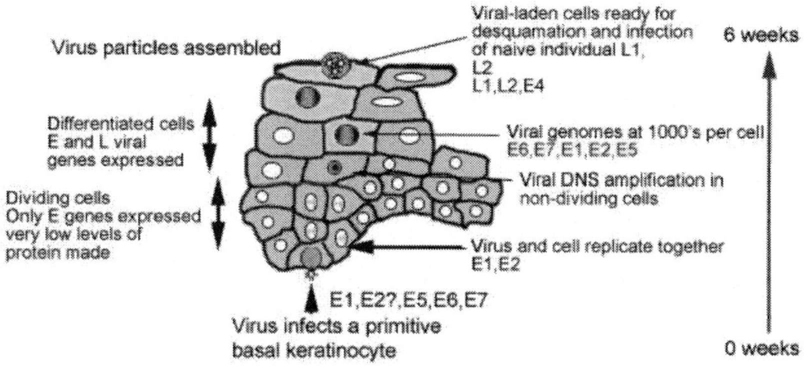

Figure 3. HPV penetrating the basal layer and released at the epithelial surface (Medscape [109]).

Although most HPVs are nonpathogenic, there are a small number of high-risk HPV serotypes, especially HPV-16 and -18, which together cause approximately 70% of all cervical cancers. In most HPV-induced cancers, the HPV episome is found clonally integrated into the cell genome in a manner that destroys or deletes the viral E2 gene (E for early) [31]. The role of the E2 protein is to bind to the HPV origin of replication, where it functions to ensure the distribution of HPV episomes to both daughter cells after cell division, and E2 also acts to regulate HPV early gene transcription. One key activity of E2 is to limit the expression of the HPV oncogenes E6 and E7, and disruption of E2 during integration into the host cell genome can lead to high constitutive levels of E6 and E7 expression [110]. E6 functions to bind and destabilize the p53 tumor suppressor [111], whereas E7 similarly binds and destabilizes the Rb tumor suppressor [112], and these two functions play a critical role in the maintenance of HPV-transformed cells. Cancers associated with HPV infection include cervical carcinoma, which is almost always HPV positive, as well as a fraction of head and neck (H&N) carcinoma and anal cancer, all of which are related to sexual transmission of HPV. Novel treatment for chemoresistant HPV-positive tumors will be important for treating recurrent disease. Of note, almost all HPV-positive H&N and anal cancers are HPV 16 positive, thus restricting the required sequence range for engineered endonuclease-based therapy.

Both HPV E6 and E7 proteins play a crucial role in HPV tumorigenesis by blocking the action of p53 and Rb, respectively [112]. Consistent with this idea, the inactivation of the E6 gene in the HPV-18-positive cervical carcinoma cell line HeLa or the HPV-16-positive cell line SiHa using Spy CRISPR/Cas has been found to result in the induction of p53 expression followed by the expression of downstream targets of this cellular transcription factor, including the CDK inhibitor p21 and several activators of apoptosis, leading to cell cycle arrest and cell death [68, 82]. Similarly, disruption of the E7 gene using CRISPR/Cas results in the increased expression of Rb, formation of Rb/E2F heterodimers, and then the induction of cellular genes that induce senescence and cell death [68, 113]. Mino et al. [81] improved the design of a ZFN-based hybrid nuclease to the single-chain FokI dimer (scFokI) and demonstrated that it inhibited HPV-18 DNA replication in transient replication assays using mammalian cells more efficiently. By linker-mediated PCR analysis, the researchers confirmed that AZP-scFokI cleaved an HPV-18 ori plasmid around its binding site in mammalian cells. These studies suggest that targeted endonucleases specific for HPV E6 and/or E7 has the potential to serve as a novel, highly specific, and effective therapy for chemoresistant HPV-16 induced anal and H&N tumors.

2.3.5. *Other viruses*

A number of other chronic virus infections are associated with serious human diseases including EBV, cytomegalovirus, HCV, Kaposi's sarcoma-associated herpes virus (KSHV), human T-cell leukemia virus type 1 (HTLV-1), and Merkel cell polyomavirus (MCPyV) (see **Tables 2** and **3**). Of these, perhaps the most relevant in relation to DNA editing technology is EBV. EBV is the etiologic agent of several cancers, including an epithelial cell tumor called nasopharyngeal carcinoma (NPC), which is highly prevalent in southern China and Southeast Asia [40, 52]. In NPC cells, EBV is found in a form of viral latency that nevertheless involves the expression of several viral nonstructural proteins and microRNAs [114]. EBV-positive NPCs share a number of characteristics with HPV-16-positive H&N cancers, and as in the latter case, the continued presence and transcription of the viral (in this case, EBV) genome is thought to be essential for tumor survival. Wang and Quake [70] used the CRISPR/Cas9 system for antiviral therapy in human cells, specifically targeting the EBV genomes of latent viral infections. Patient-derived Burkitt's lymphoma cells with latent EBV infection showed significant proliferation arrest and decrease in viral load after treatment with specific CRISPR/Cas9 vector targeting the viral genome. It seems likely that NPC cells would be excellent targets for transduction in vivo using Sau Cas9/sgRNA-based AAV vectors specific for the EBV genome.

Progressive multifocal leukoencephalopathy (PML) is a fatal demyelinating disease of the central nervous system caused by human polyomavirus JC (JCV) reactivation. JCV replicates in oligodendrocytes, the myelin-producing cells in the brain. Previously a rare disease seen in patients with lymphoproliferative and myeloproliferative disorders, PML is now seen more frequently in HIV-1-positive/AIDS patients and patients undergoing immunomodulatory therapy due for rheumatological/autoimmune disorders [83]. At this time, there is no cure for PML, and in most cases, disease progression leads to death within 2 years. The JCV genome is a small circular double-stranded DNA that includes coding sequences for the viral early protein, T-antigen, which is critical for directing viral reactivation and lytic infection. Wollebo et al. [83] applied CRISPR/Cas9 system to introduce mutations in the viral genome to inactivate the gene encoding T-antigen and inhibit viral replication. Transient or conditional expression of Cas9 and gRNAs specifically targets the N-terminal region of T-antigen on integrated genetic and functional studies. The mutation introduced interferes with the expression and function of the viral protein, suppressing viral replication in vitro [83]. There was no off-target effect of the JCV-specific CRISPR/Cas9 editing apparatus observed. These studies provide the first evidence for the employment of a gene editing strategy as a promising tool for the elimination of the JCV genome and a potential cure for PML.

HTLV-1, which causes adult T-cell leukemia (ATL) in humans, establishes a lifelong latent infection. Current therapies are not very effective against HTLV-1-associated disorders. In a proof-of-concept study, Tanaka et al. [80] developed a targeted endonuclease based on zinc finger nuclease (ZFN) that specifically recognized a conserved region of HTLV-1 long terminal repeat (LTR) and introduced it into various HTLV-1-positive human T-cell lines, including HTLV-1-transformed and ATL-derived cell lines [80]. ZFN disrupted the promoter function of HTLV-1 LTR and specifically killed HTLV-1-infected cells. The researchers showed the first evidence of the removal of the proviral genome from HTLV-1-infected cells. The therapeutic effect of ZFN was confirmed in an in vivo model of ATL. This strategy may form the basis of a therapy that can eradicate HTLV-1 infection, and similar approaches can be used to target other malignancy-associated viruses.

Using DNA editing technology to inactivate or mutate DNA virus genomes

Virus	Human disease	Type	Target	Cell/animal/ human study	Delivery method	Reference
Endogenous retroviruses (ERV)	Unknown	ZFN, CRISPR	Pol	ATCC CRL-33, PK15 Pig ATCC CRL-1573, human	Transfection	Semaan et al. [69], Mali et al. [6], Cong et al. [5]
Epstein–Barr virus (EBV)	Burkitt's lymphoma, NPC, non-Hodgkin's lymphomas, oral hairy leukoplakia, lymphoproliferative disease	CRISPR	EBNA1	Raji cells, human	Transfection	Wang et al. [70]
Hepatitis B virus (HBV)	Liver cirrhosis, HCC	ZFN	Pol, Core, X	Huh7 cells, HepAd38 cells	Transfection, AAV vectors	Cradick et al. [71], Weber et al. [72]
		TALEN	Surface, Core, Pol	HepG2.2.15, mouse	Transfection, HDI	Bloom K et al. [73]
		TALEN	Core	Huh7 cells	Transfection	Chen et al. [74] Lin et al. [75], Seeger et al. [76],
		CRISPR	Core, X	HepG2/NTCP cells	Lentiviral vectors	Kennedy et al. [77], Dong et al. [78]
Herpes simplex type 1 (HSV-1)	Core sores, encephalitis, pneumonitis, hepatitis	Homing endonuclease (HE) HSV1m5	HSV-1 gene UL19	Primary human fibroblast	AAV vectors	Aubert et al. [79]

Using DNA editing technology to inactivate or mutate DNA virus genomes						
Virus	**Human disease**	**Type**	**Target**	**Cell/animal/ human study**	**Delivery method**	**Reference**
Human T cell leukemia virus type 1 (HTLV1)	Adult T cell leukemia (2–6% of carriers), tropical spastic paraparesis, myelopathy, uveitis, dermatitis	ZFN	HTLV-1 LTR	T-cell and ATL-derived cell lines		Tanaka et al. [80]
Human Papilloma virus	Papilloma, cervical and other mucosal carcinoma	ZFN	HPV-18 ori	293H cells	Transfection	Mino et al. [81]
		CRISPR	E6, E7 ORFs	SiHa cells, mice	Transfection	Zhen et al. [82]
Polyomavirus JC	Progressive multifocal leukoencephalopathy	CRISPR	JCV T-antigen gene	HJC-2, Hamster	Transfection	Wollebo et al. [83]

Table 3. Summary of studies using DNA editing technology to inactivate or mutate DNA virus genomes.

3. Conclusion

The development of effective gene editing technologies has the potential to lead to the global identification of almost all cellular factors that regulate virus replication in culture, leading to a wealth of new insights into viral molecular biology and producing numerous potential targets for antiviral drug development. If such screens can indeed identify cellular factors that are required for virus replication but entirely dispensable for the health of the adult target organism, then it might be possible to also treat viral infections, including RNA virus infections, via the localized ablation of a cellular gene as described above for direct targeting of DNA virus genomes. A similar approach has previously been reported using ZFNs to inactivate the HIV-1 coreceptor CCR5, to prevent the infection of $CD4^+$ T cells, by transduction of hematopoietic stem cells followed by an analysis of the production of HIV-1-resistant $CD4^+$ human T cells in engrafted immunodeficient mice [115, 116]. The challenges remain in applying engineered nucleases in the clinic setting, including immunogenicity of transduced proteins, with repeated application, delivery of the genes into the correct tissue/cell types, side effects of off-target mutagenesis, limitations in the design of the nuclease target sites, ethical issues, including tumorigenicity, undesired integration of nucleases or donor templates, and the germline transmission of the modified genome. One interesting and novel approach is to deliver engineered nuclease using modified mRNA, which is nonintegrating and provides a transient pulse of protein expression, as an alternative to traditional viral vectors. A team of researchers had recently applied this nuclease-encoding, chemically modified (nec) mRNA to deliver site-specific nucleases in a transgenic mouse model of SP-B deficiency, resulting in successful site-specific genome editing in vivo [117, 118]. Although several technical challenges and uncertainties remain, the promise of using gene correction technology to study and treat chronic viral infections is tremendous. Further advances in understanding and improvements in technology will open the next era of therapy against currently difficult-to-treat viral diseases.

Author details

Guan-Huei Lee[1,2*] and Myo Myint Aung[1]

*Address all correspondence to: leeguanhuei@gmail.com

1 Department of Medicine, National University Health System, Singapore, Singapore

2 Yong Loo Lin School of Medicine, National University of Singapore, Singapore, Singapore

References

[1] Klug A. The discovery of zinc fingers and their applications in gene regulation and genome manipulation. Annu Rev Biochem. 2010;79:213–231.

[2] Gaj T, Gersbach CA, Barbas CF 3rd. ZFN, TALEN, and CRISPR/Cas-based methods for genome engineering. Trends Biotechnol. 2013;31:397–405.

[3] Christian M, Cermak T, Doyle EL, et al. Targeting DNA double-strand breaks with TAL effector nucleases. Genetics. 2010;186:757–761.

[4] Miller JC, Tan S, Qiao G, et al. A TALE nuclease architecture for efficient genome editing. Nat Biotechnol. 2011;29:143–148.

[5] Barrangou R, Marraffini LA. CRISPR-Cas systems: Prokaryotes upgrade to adaptive immunity. Mol Cell. 2014 Apr 24;54(2):234–44.

[6] Hsu PD, Scott DA, Weinstein JA, Ran FA, Konermann S, Agarwala V, Li Y, Fine EJ, Wu X, Shalem O, et al. DNA targeting specificity of RNA-guided Cas9 nucleases. Nat. Biotechnol. 2013, 31, 827–832.

[7] Cong L, Ran FA, Cox D, et al. Multiplex genome engineering using CRISPR/Cas systems. Science. 2013;339:819–823.

[8] Mali P, Yang L, Esvelt KM, et al. RNA-guided human genome engineering via Cas9. Science. 2013;339:823–826.

[9] Segal DJ, Meckler JF. Genome engineering at the dawn of the golden age. Annu Rev Genom Hum Genet. 2013;14:135–158.

[10] Kim H, Kim JS. A guide to genome engineering with programmable nucleases. Nat Rev. 2014;15:321–334.

[11] Ousterout DG, et al. Reading frame correction by targeted genome editing restores dystrophin expression in cells from Duchenne muscular dystrophy patients. Mol Ther. 2013;21:1718–1726.

[12] Jiang J, et al. Translating dosage compensation to trisomy 21. Nature. 2013;500:296–300.

[13] Zou J, et al. Oxidase-deficient neutrophils from X-linked chronic granulomatous disease i PS cells: functional correction by zinc finger nuclease-mediated safe harbor targeting. Blood. 2011;117:5561–5572.

[14] Schwank G, et al. Functional repair of CFTR by CRISPR/Cas9 in intestinal stem cell organoids of cystic fibrosis patients. Cell Stem Cell. 2013;13:653–658.

[15] Yusa K, et al. Targeted gene correction of α1-antitrypsin deficiency in induced pluripotent stem cells. Nature. 2011;478:391–394.

[16] Sebastiano V, et al. In situ genetic correction of the sickle cell anemia mutation in human induced pluripotent stem cells using engineered zinc finger nucleases. Stem Cells. 2011;29:1717–1726.

[17] Osborn MJ, et al. TALEN-based gene correction for epidermolysis bullosa. Mol Ther. 2013;21:1151–1159.

[18] Virgin HW, Wherry EJ, Ahmed R. Redefining chronic viral infection. Cell. 2009;138:30–50.

[19] Mc Govern BH. The epidemiology, natural history and prevention of hepatitis B: implications of HIV coinfection. Antivir Ther. 2007;12:H3–H13.

[20] Rehermann B, Nascimbeni M. Immunology of hepatitis B virus and hepatitis C virus infection. Nat Rev Immunol. 2005;5:215–229.

[21] Lemon SM, Walker CM, Alter MJ, et al. Hepatitis C virus. In: Knipe D, Howley P, editors. Fields Virology. Philadelphia: Lippincott, Williams & Wilkins; 2007:1253–1304.

[22] UNAIDS. 2008 report on the global AIDS epidemic; 2008. Available at: http://www.unaids.org/en/KnowledgeCentre/HIVData/GlobalReport/2008/2008_Global_report.asp.

[23] Kuritzkes DR, Walter BD. HIV-1: pathogenesis, clinical manifestations, and treatment. In: Knipe D, Howley P, editors. Fields Virology. Philadelphia: Lippincott, Williams & Wilkins; 2007:2187–2214.

[24] Taylor JM, Farci P, Purcell RH. Hepatitis D (Delta) virus. In: Knipe D, Howley P, editors. Fields Virology. Philadelphia: Lippincott, Williams & Wilkins; 2007:3031–3046.

[25] Matsuoka M, Jeang KT. Human T-cell leukaemia virus type 1 (HTLV-1) infectivity and cellular transformation. Nat Rev Cancer. 2007;7:270–280.

[26] Lairmore MD, Franchini G. Human T-cell. In: Knipe D, Howley P, editors. Fields Virology. Philadelphia: Lippincott, Williams & Wilkins; 2007:2071–2103.

[27] Urisman A, Molinaro RJ, Fischer N, et al. Identification of a novel gammaretrovirus in prostate tumors of patients homozygous for R462Q RNASEL variant. PLo S Pathog. 2006;2:e25.

[28] Dong B, Kim S, Hong S, et al. An infectious retrovirus susceptible to an IFN antiviral pathway from human prostate tumors. Proc Natl Acad Sci USA. 2007;104:1655–1660.

[29] Hobman T, Chantler J. Rubella virus. In: Knipe D, Howley P, editors. Fields Virology. Philadelphia: Lippincott, Williams & Wilkins; 2007:1069–1100.

[30] Berns K, Parrish CR. Parvoviridae. In: Knipe D, Howley P, editors. Fields Virology. Philadelphia: Lippincott, Williams & Wilkins; 2008:2437–2466.

[31] Norja P, Eis-Hubinger AM, Soderlund-Venermo M, et al. Rapid sequence change and geographical spread of human parvovirus B19: comparison of B19 virus evolution in acute and persistent infections. J Virol. 2008;82:6427–6433.

[32] Leggatt GR, Frazer IH. HPV vaccines: the beginning of the end for cervical cancer. Curr Opin Immunol. 2007;19:232–238.

[33] Howley P, Lowy DR. Papillomaviruses. In: Knipe D, Howley P, editors. Fields Virology. Philadelphia: Lippincott, Williams & Wilkins; 2007:2297–2354.

[34] Griffin DE. Measles virus. In: Knipe D, Howley P, editors. Fields Virology. Philadelphia: Lippincott, Williams & Wilkins; 2007:1551–1585.

[35] Chapman NM, Kim KS. Persistent coxsackievirus infection: enterovirus persistence in chronic myocarditis and dilated cardiomyopathy. Curr Top Microbiol Immunol. 2008;323:275–292.

[36] Whitton JL, Feuer R. Myocarditis, microbes and autoimmunity. Autoimmunity. 2004;37:375–386.

[37] Straus SE. Introduction to herpesviridae. In Mandel G, Bennett JE, Dolin R, editors. Mandell, Douglas, and Bennett's Principles and Practice of Infectious Diseases. Philadelphia: Churchill Livingstone; 2000:1557–1564.

[38] Yamanishi K, Mori Y, Pellett PE. Human herpesviruses 6 and 7. In: Knipe D, Howley P, editors. Fields Virology. Philadelphia: Lippincott, Williams & Wilkins; 2007:2819–2845.

[39] Zerboni L, Arvin AM. The pathogenesis of varicella-zoster virus neurotropism and infection. In Reiss CS, editor. Neurotropic Viral Infections. Cambridge: Cambridge University Press; 2008:225–250.

[40] Straus SE. Introduction to herpesviridae. In Mandel G, Bennett JE, Dolin R, editors. Mandell, Douglas, and Bennett's Principles and Practice of Infectious Diseases. Philadelphia: Churchill Livingstone; 2000:1557–1564.

[41] Mocarski ES, Shenk T, Pass RF. Cytomegaloviruses. In: Knipe D, Howley P, editors. Fields Virology. Philadelphia: Lippincott, Williams & Wilkins; 2007:2701–2772.

[42] Rickinson A, Kieff E. Epstein-Barr virus. In Knipe D, Griffin DE, Lamb RA, et al., editors. Fields Virology. Philadelphia: Lippincott, Williams & Wilkins; 2007:2655–2700.

[43] Kieff E, Rickinson A. Epstein-Barr virus and its replication. In: Knipe D, Howley P, editors. Fields Virology. Philadelphia: Lippincott, Williams & Wilkins; 2007:2605–2654.

[44] Zur HH. Novel human polyomaviruses—re-emergence of a well known virus family as possible human carcinogens. Int J Cancer. 2008;123:247–250.

[45] Garnett CT, Erdman D, Xu W, et al. Prevalence and quantitation of species C adenovirus DNA in human mucosal lymphocytes. J Virol. 2002;76:10608–10616.

[46] Wold WS, Horwitz MS. Adenoviruses. In: Knipe D, Howley P, editors. Fields Virology. Philadelphia: Lippincott, Williams & Wilkins; 2008:2395–2425.

[47] Ganem D. KSHV infection and the pathogenesis of Kaposi's sarcoma. Annu Rev Pathol. 2006;1:273–296.

[48] Guidotti LG, Chisari FV. Immunobiology and pathogenesis of viral hepatitis. Annu Rev Pathol. 2006;1:23–61.

[49] Chun TW, Stuyver L, Mizell SB, et al. Presence of an inducible HIV-1 latent reservoir during highly active antiretroviral therapy. Proc Natl Acad Sci USA. 1997;94:13193–13197.

[50] Finzi D, Hermankova M, Pierson T, et al. Identification of a reservoir for HIV-1 in patients on highly active antiretroviral therapy. Science. 1997;278:1295–1300.

[51] Wong JK, Hezareh M, Gunthard HF, et al. Recovery of replication-competent HIV despite prolonged suppression of plasma viremia. Science. 1997;278:1291–1295.

[52] Boldogh I, Albrecht T, Porter DD. Persistent viral infections. In: Baron S, editor. Medical Microbiology. 4th ed. Galveston, TX: University of Texas Medical Branch at Galveston; 1996:Chapter 46.

[53] Schiffer JT, Aubert M, Weber ND, et al. Targeted DNA mutagenesis for the cure of chronic viral infections. J Virol. 2012;86:8920–8936.

[54] Kennedy EM, Cullen BR. Bacterial CRISPR/Cas DNA endonucleases: a revolutionary technology that could dramatically impact viral research and treatment. Virology. 2015;479–480:213–220.

[55] Malim MH, Bieniasz PD. HIV restriction factors and mechanisms of evasion. Cold Spring Harb Perspect Med. 2012;2:a006940.

[56] Shalem O, Sanjana NE, Hartenian E, et al. Genome-scale CRISPR-Cas9 knockout screening in human cells. Science. 2014;343:84–87.

[57] Wang T, Wei JJ, Sabatini DM, et al. Genetic screens in human cells using the CRISPR-Cas9 system. Science. 2014;343:80–84.

[58] Zhou Y, Zhu S, Cai C, et al. High-throughput screening of a CRISPR/Cas9 library for functional genomics in human cells. Nature. 2014;509:487–491.

[59] Menger L, Gouble A, Marzolini MA et al. TALEN-mediated genetic inactivation of the glucocorticoid receptor in cytomegalovirus-specific T cells. Blood. 2015;126:2781–2789.

[60] Sung PS, Murayama A, Kang W, et al. Hepatitis C virus entry is impaired by claudin-1 downregulation in diacylglycerol acyltransferase-1-deficient cells. J Virol. 2014;88:9233–9244.

[61] Ren Q, Li C, Yuan P et al. A dual-reporter system for real-time monitoring and high-throughput CRISPR/Cas9 library screening of the hepatitis C virus. Sci Rep. 2015;5:8865.

[62] Domingues P, Bamford CG, Boutell C, et al. Inhibition of hepatitis C virus RNA replication by ISG15 does not require its conjugation to protein substrates by the HERC5 E3 ligase. J Gen Virol. 2015.

[63] Johnson KE, Bottero V, Flaherty S, et al. IFI16 restricts HSV-1 replication by accumulating on the HSV-1 genome, repressing HSV-1 gene expression, and directly or indirectly modulating histone modifications. PLo S Pathog. 2014;10:e1004503.

[64] Turner EM, Brown RS, Laudermilch E, et al. The torsin activator LULL1 is required for efficient growth of herpes simplex virus 1. J Virol. 2015;89:8444–8452.

[65] Perez EE, Wang J, Miller JC, et al. Establishment of HIV-1 resistance in $CD4^+$ T cells by genome editing using zinc-finger nucleases. Nat Biotechnol. 2008;26:808–816.

[66] Tebas P, Stein D, Tang WW, et al. Gene editing of CCR5 in autologous CD4 T cells of persons infected with HIV. N Engl J Med. 2014;370:901–910.

[67] Li C, Guan X, Du T, et al. Inhibition of HIV-1 infection of primary $CD4^+$ T-cells by gene editing of CCR5 using adenovirus-delivered CRISPR/Cas9. J Gen Virol. 2015;96:2381–2393.

[68] Avey D, Tepper S, Li W, et al. Phosphoproteomic analysis of KSHV-infected cells reveals roles of ORF45-activated RSK during lytic replication. PLo S Pathog. 2015;11:e1004993.

[69] Hu Z, Ding W, Zhu D, et al. TALEN-mediated targeting of HPV oncogenes ameliorates HPV-related cervical malignancy. J Clin Invest. 2015;125:425–436.

[70] Kennedy EM, Kornepati AV, Goldstein M, et al. Inactivation of the human papillomavirus E6 or E7 gene in cervical carcinoma cells by using a bacterial CRISPR/Cas RNA-guided endonuclease. J Virol. 2014;88:11965–11972.

[71] Semaan M, Ivanusic D, Denner J. Cytotoxic effects during knock out of multiple porcine endogenous retrovirus (PERV) sequences in the pig genome by zinc finger nucleases (ZFN). PLoS One. 2015;10:e0122059.

[72] Wang J, Quake SR. RNA-guided endonuclease provides a therapeutic strategy to cure latent herpesviridae infection. Proc Natl Acad Sci USA. 2014;111:13157–13162.

[73] Cradick TJ, Keck K, Bradshaw S, et al. Zinc-finger nucleases as a novel therapeutic strategy for targeting hepatitis B virus DNAs. Mol Ther. 2010;18:947–954.

[74] Weber ND, Stone D, Sedlak RH, et al. AAV-mediated delivery of zinc finger nucleases targeting hepatitis B virus inhibits active replication. PLo S ONE. 2014;9:e97579.

[75] Bloom K, Ely A, Mussolino C et al. Inactivation of hepatitis B virus replication in cultured cells and in vivo with engineered transcription activator-like effector nucleases. Mol Ther. 2013;21:1889–1897.

[76] Chen J, Zhang W, Lin J, et al. An efficient antiviral strategy for targeting hepatitis B virus genome using transcription activator-like effector nucleases. Mol Ther. 2014;22:303–311.

[77] Lin SR, Yang HC, Kuo YT, et al. The CRISPR-cas9 system facilitates clearance of the intrahepatic HBV templates in vivo. Mol Ther. Nucleic Acids 2014;3:e186.

[78] Seeger C, Sohn JA. Targeting hepatitis B virus with CRISPR-cas9. Mol Ther. Nucleic Acids 2014;3:e216.

[79] Kennedy EM, Bassit LC, Mueller H, et al. Suppression of hepatitis B virus DNA accumulation in chronically infected cells using a bacterial CRISPR-Cas RNA-guided DNA endonuclease. Virology. 2015;476:196–205.

[80] Dong C, Qu L, Wang H, et al. Targeting hepatitis B virus cccDNA by CRISPR-cas9 nuclease efficiently inhibits viral replication. Antivir Res. 2015;118:110–117.

[81] Aubert M, Boyle NM, Stone D, et al. In vitro inactivation of latent HSV by targeted mutagenesis using an HSV-specific homing endonuclease. Mol Ther Nucleic Acids. 2014;3:e146.

[82] Tanaka A, Takeda S, Kariya R, et al. A novel therapeutic molecule against HTLV-1 infection targeting provirus. Leukemia. 2013;27:1621–7.

[83] Mino T, Mori T, Aoyama Y, et al. Inhibition of DNA replication of human papillomavirus by using zinc finger-single-chain Fok I dimer hybrid. Mol Biotechnol. 2014;56:731–737.

[84] Zhen S, Hua L, Takahashi Y, et al. In vitro and in vivo growth suppression of human papillomavirus 16-positive cervical cancer cells by CRISPR/Cas9. Biochem Biophys Res Commun. 2014;450:1422–1426.

[85] Wollebo HS, Bellizzi A, Kaminski R, et al. CRISPR/Cas9 system as an agent for eliminating polyomavirus JC infection. PLoS One. 2015;10:e0136046.

[86] Manjunath N, Yi G, Dang Y, et al. Newer gene editing technologies toward HIV gene therapy. Viruses. 2013;5:2748–2766.

[87] Xing S, SilicianoRF. Targeting HIV latency: pharmacologic strategies toward eradication. Drug Discov Today. 2013;18:541–551.

[88] Ebina H, Misawa N, Kanemura Y, et al. Harnessing the CRISPR/Cas9 system to disrupt latent HIV-1 provirus. Sci Rep. 2013;3:2510.

[89] Hu W, Kaminski R, Yang F, et al. RNA-directed gene editing specifically eradicates latent and prevents new HIV-1 infection. Proc Natl Acad Sci USA. 2014;111:11461–11466.

[90] Alkhatib G, Combadiere C, Broder CC, et al. CC CKR5: a RANTES, MIP-1alpha, MIP-1beta receptor as a fusion cofactor for macrophage-tropic HIV-1. Science. 1996; 272:1955–1958.

[91] Deng H, Liu R, Ellmeier W, et al. Identification of a major co-receptor for primary isolates of HIV-1. Nature. 1996;381:661–666.

[92] Liu R, Paxton WA, Choe S, et al. Homozygous defect in HIV-1 coreceptor accounts for resistance of some multiply-exposed individuals to HIV-1 infection. Cell. 1996;86:367–377.

[93] Samson M, Libert F, Doranz BJ, et al. Resistance to HIV-1 infection in Caucasian individuals bearing mutant alleles of the CCR-5 chemokine receptor gene. Nature. 1996;382:722–725.

[94] Eugen-Olsen J, Iversen AK, Garred P, et al. Heterozygosity for a deletion in the CKR-5 gene leads to prolonged AIDS-free survival and slower CD4 T-cell decline in a cohort of HIV-seropositive individuals. AIDS. 1997;11:305–310.

[95] Cohen OJ, Vaccarezza M, Lam GK, et al. Heterozygosity for a defective gene for CC chemokine receptor 5 is not the sole determinant for the immunologic and virologic phenotype of HIV-infected long-term nonprogressors. J Clin Invest. 1997; 100:1581–1589.

[96] Gulick RM, Lalezari J, Goodrich J, et al. Maraviroc for previously treated patients with R5 HIV-1 infection. N Engl J Med 2008;359:1429–1441.

[97] Hütter G, Nowak D, Mossner M, et al. Long-term control of HIV by CCR5 delta32/delta32 stem-cell transplantation. N Engl J Med. 2009;360:692–698.

[98] Allers K, Hütter G, Hofmann J, et al. Evidence for the cure of HIV infection by CCR5Δ32/Δ32 stem cell transplantation. Blood. 2011;117:2791–2799.

[99] Deeks SG, Mc Cune JM. Can HIV be cured with stem cell therapy? Nat Biotechnol. 2010;28:807–810.

[100] Komatsu H. Hepatitis B virus: where do we stand and what is the next step for eradication? World J Gastroenterol. 2014;20:8998–9016.

[101] Werle-Lapostolle B, Bowden S, Locarnini S, et al. Persistence of cccDNA during the natural history of chronic hepatitis B and decline during adefovir dipivoxil therapy. Gastroenterology. 2004;126:1750–1758.

[102] Phyo WW, Soh AY, Lim SG, Lee GH. Search for a cure for chronic hepatitis B infection: How close are we? World J Hepatol. 2015;7:1272–1281.

[103] Zimmerman KA, Fischer KP, Joyce MA, et al. Zinc finger proteins designed to specifically target duck hepatitis B virus covalently closed circular DNA inhibit viral transcription in tissue culture. J Virol. 2008;82:8013–8021.

[104] Kew MC. Hepatitis B virus X protein in the pathogenesis of hepatitis B virus-induced hepatocellular carcinoma. J Gastroenterol Hepatol. 2011;26:144–152.

[105] Zhao X, Zhao Z, Guo J, et al. Creation of a six-fingered artificial transcription factor that represses the hepatitis B virus HBx gene integrated into a human hepatocellular carcinoma cell line. J Biomol Screen. 2013;18:378–387.

[106] Qi Z, Li G, Hu H, et al. Recombinant covalently closed circular hepatitis B virus DNA induces prolonged viral persistence in immunocompetent mice. J Virol. 2014;88:8045–8056.

[107] Lisowski L, Dane AP, Chu K, et al. Selection and evaluation of clinically relevant AAV variants in a xenograft liver model. Nature. 2014;506:382–386.

[108] Roizman B, Knipe DM, Whitley RJ. Herpes simplex viruses. In: Knipe DM, Howley PM, editors, Fields Virology. 5th ed. Philadelphia: Lippincott Williams & Wilkins; 2007:2501–2601.

[109] Umbach JL, Kramer MF, Jurak I, et al. MicroRNAs expressed by herpes simplex virus 1 during latent infection regulate viral mRNAs. Nature. 2008;454:780–783.

[110] Liesegang TJ. Herpes simplex virus epidemiology and ocular importance. Cornea. 2001;20:1–13.

[111] Medscape. Human papillomavirus—pathophysiology [Internet]; 2015. Available at: http://emedicine.medscape.com/article/219110-overview#a3. Accessed 2016-01-02.

[112] Thierry F, Howley PM. Functional analysis of E2-mediated repression of the HPV18 P105 promoter. New Biol. 1991;3:90–100.

[113] Scheffner M, Werness BA, Huibregtse JM, et al. The E6 oncoprotein encoded by human papillomavirus types 16 and 18 promotes the degradation of p53. Cell. 1990;63:1129–1136.

[114] Mighty KK, Laimins LA. The role of human papillomaviruses in oncogenesis. Recent Results Cancer Res. 2014;193:135–148.

[115] Hu Z, Yu L, Zhu D, et al. Disruption of HPV16-E7 by CRISPR/Cas system induces apoptosis and growth inhibition in HPV16 positive human cervical cancer cells. Biomed Res Int. 2014;2014:612823.

[116] Tsao SW, Tsang CM, Pang PS, et al. The biology of EBV infection in human epithelial cells. Semin Cancer Biol. 2012;22:137–143.

[117] Holt N, Wang J, Kim K, et al. Human hematopoietic stem/progenitor cells modified by zinc-finger nucleases targeted to CCR5 control HIV-1 in vivo. Nat Biotechnol. 2010;28:839–847.

[118] Li L, Krymskaya L, Wang J, et al. Genomic editing of the HIV-1 coreceptor CCR5 in adult hematopoietic stem and progenitor cells using zinc finger nucleases. Mol Ther. 2013;21:1259–1269.

[119] Kormann MS, Hasenpusch G, Aneja MK, et al. Expression of therapeutic proteins after delivery of chemically modified mRNA in mice. Nat Biotechnol. 2011;29:154–157.

[120] Mahiny AJ, Dewerth A, Mays LE, et al. In vivo genome editing using nuclease-encoding mRNA corrects SP-B deficiency. Nat Biotechnol. 2015;33:584–586.

Chapter 5

Gene Editing in Adult Hematopoietic Stem Cells

Sergio López-Manzaneda, Sara Fañanas-Baquero,
Virginia Nieto-Romero,
Francisco-Jose Roman-Rodríguez,
Maria Fernandez-Garcia, Maria J. Pino-Barrio,
Fatima Rodriguez-Fornes, Begoña Diez-Cabezas,
Maria Garcia-Bravo, Susana Navarro,
Oscar Quintana-Bustamante and Jose C. Segovia

Additional information is available at the end of the chapter

http://dx.doi.org/10.5772/62383

Abstract

Over the last years, an important development has allowed the scientific community to address a precise and accurate modification of the genome. The first probe of concept appeared with the design and use of engineered zinc-finger nucleases (ZFNs), which was expanded later on with the discovery and engineering of meganucleases and transcription activator-like effector nucleases (TALENs) and finally democratized and made easily available to the whole scientific community with the discovery of the clustered regularly interspaced short palindromic repeats (CRISPR)/Cas9 nuclease combination technology. The availability of these tools has allowed a precise gene editing, such as knockout of a specific gene or the correction of a defective gene by means of homologous recombination (HR), taking advantage of the endogenous cell repair machinery. This process was already known and used but was inefficient—efficiency that has been increased more than 100-fold with the addition of the mentioned specific nucleases to the process. Apart from the proper design of the nucleases to recognize and cut the selected site in the cell genome, two main goals need to be adequately addressed to optimize its function: the delivery of the tools into the desired cells and the selection of those where the gene editing process has occurred correctly. Both steps can be easily solved when the source of cells is extensive or can be expanded and manipulated *in vitro* extensively, such as immortalized cell lines or pluripotent stem cells (embryonic stem cells and induced pluripotent stem cells). However, both steps are critical in the case of primary cells, such as the hematopoietic stem cells (HSCs). HSCs are a rare cell population present in the bone marrow (BM) of higher mammals, and it is the responsible for the maintenance and replenishment of all hematopoietic cells for the lifespan of the animals by means of two fundamental properties: self-renewal and multipotency. HSC population is then the ideal target for the correction of hematopoietic genetic diseases and also for the knockout of the responsible genes to *in vitro* and *in vivo* model those hematopoietic diseases. This rare population cannot be expanded and its *in vitro* manipulation and culture negatively affects their fundamental properties of self-renewal and multipotency. These factors challenge the application of gene editing to HSCs. Important efforts are now ongoing trying to optimize the protocols of gene delivery and selection for HSCs. This chapter will review and discuss how researchers are trying to solve them, all attempts that are ongoing and the potential application of the technology to the patients affected with hematopoietic genetic diseases.

Keywords: hematopoietic stem cells, hematopoiesis, genetic disease, gene therapy, gene editing, nucleases, zinc finger, TALEN, meganucleases, CRISPR/Cas9

1. Introduction

Adult hematopoietic stem cells (HSCs) are a rare population of cells that are present in the bone marrow (BM) and are the responsible for the generation of all mature blood cells, including erythrocytes, platelets, and immune cells (1). This population represents less than 0.1% of the total BM and is the only one capable of self-renewal, being responsible for the maintenance of the lifelong hematopoiesis. Since their definition at the beginning of the last century (2), an extensive knowledge of this population has been accumulated, being probably the best known adult stem cell. HSCs can be phenotypically and functionally characterized and can be purified and even manipulated *in vitro*. Once transplanted back into the animal or human body, HSCs are capable of regenerating the whole hematopoietic system of the organism (1,3). These capabilities established the possibility of BM transplantation as a therapeutic strategy for the treatment of hematopoietic diseases, either inherited or acquired. BM transplantation can be autologous, when the HSCs come from the same person, or allogeneic, when they come from a healthy donor. Allogeneic BM transplant is the only curative treatment for oncogenic diseases and for inherited (genetic) diseases, providing healthy HSCs. However, the immune reaction of the transplanted cells against the recipient [graft-versus-host disease (GVHD)] and the availability of compatible BM donors reduce dramatically its applicability and increase the severe adverse effects of the treatment (4). To definitively solve this problem, the genetic modification of the autologous HSCs to restore the genetic defect or to provide them with new capabilities, the so-called gene therapy, appeared as an additional therapeutic option avoiding GVHD and providing repaired/cured cells.

The idea of gene therapy appeared up in the middle of the 20th century, when the use of modified viruses as vectors to introduce the desired genetic material into the desired cells was already mentioned and demonstrated as a proof-of-concept (5). Since then, gene therapy has been playing in a roller coaster with tremendous improvements and successes but also with serious problems that have been dramatic in some instances (6). Finally, gene therapy is nowadays established as a real therapeutic option for a number of hematopoietic inherited diseases, such as immunodeficiencies or hemoglobinopathies (7). The use of self-inactivating lentiviral vectors, the improvement of the *in vitro* culture conditions of the human HSCs, and the better knowledge of HSC engraftment in the recipient has contributed to this reality (8). Nevertheless, HSC gene therapy still has some drawbacks that needed to be addressed to make this strategy safer. The most important one is the potential insertional mutagenesis that could derive in an oncologic problem due to the inability to control the insertion site of the genetic material using the present available tools (8).

Precise gene editing using the gene repair cellular machinery appeared as the next step to follow. Widely used in the generation of animal models, mainly fish and mouse, its application as a therapeutic option in humans was far to be considered due to the low efficacy and the difficulty to select those cells where the desired gene edition had taken place. The discovery of enzymes and biological systems able to cut in one precise and specific sequence of the genome (nucleases and nuclease systems) and activate the endogenous DNA repair pathways has clearly approached gene editing to be considered as a potential therapeutic approach in the near future. First tested for their therapeutic potential for hematopoietic genetic diseases in induced pluripotent stem cells (hiPSCs) (9), the principal bottleneck in these experiments was the generation of functional HSCs from the edited hiPSCs. Thus, an increasing number of laboratories, including ours, are now adapting the gene editing protocols to adult HSCs. Important steps, such as the delivery of the gene editing tools, improving protocols to maintain HSCs *in vitro* to allow their proper edition, and improving systems to select properly edited cells, are important barriers that need to be overcome to definitely apply gene edition-based gene therapy to the clinics. Some of the steps that are now being taken are explained below.

2. Hematopoiesis and hematopoietic stem cells

The hematopoietic system is responsible for the production and maturation of the different components of the blood in a process called hematopoiesis. Blood is one of the most active regenerative tissues, with approximately 1 trillion mature cells arising daily in adult human BM (3). The hematopoiesis is a highly hierarchically organized structure in which all types of mature blood cells are generated from a small subpopulation of undifferentiated cells allocated in the BM. The hematopoietic system is divided in two main different lineages: lymphoid and myeloid. The lymphoid lineage gives rise to the T cells, B cells, and natural killer (NK) cells, whereas the myeloid lineage formed granulocytes, monocytes, erythrocytes, and megakaryocytes/platelets (1). Different organs and tissues are implicated in the production and homeostasis of the hematopoietic systems, such as blood, BM, liver, spleen, thymus, and lymph nodes.

HSCs are the most primitive cells in the hematopoietic system and support all the hematopoiesis. They are characterized by their capability to self-renew and to produce all types of mature blood functional cells. These cells persist throughout adult life, maintaining their hematopoiesis (1). Hematopoietic committed progenitor cells, the next step in the hematopoietic differentiation process, arise from the HSCs. These cells have a higher proliferative activity than HSCs, a limited self-renewal capacity and the ability to differentiate a limited number of mature cell types. Finally, mature cells complete the pyramid structure of the hematopoiesis. These cells are characterized by a recognizable morphology, full functional capabilities, and a low or null capacity of proliferation and self-renewal (10).

The study of human HSCs began with the identification of colony-forming progenitors using *in vitro* CFU-C assays (11–14). Nowadays, human HSC capacity is usually tested in xenogeneic transplants, in which human HSCs are transplanted in immunodeficient mouse strains that have a permissive microenvironment for the human HSC engraftment (15). The first humanized mouse model was the severe combined immune-deficient (SCID) mouse lacking B and T cells (16,17). Then, different mouse models were developed based on their ability to efficiently support high levels of all lineages of human engraftment (18–20). The CD34 marker was identified as a relevant indicator of human hematopoietic progenitors, which includes HSCs (21,22); nevertheless, due to the variability in their repopulation ability, several subpopulations of HSCs were identified in xenograft models, some subpopulation with a transient and limited ability to support the hematopoietic system, such as short-term HSCs (ST-HSCs), multipotent progenitors (MPP), committed progenitors, or long-term HSCs (LT-HSCs), which are able to maintain a durable hematopoiesis (23,24).

Only 1 in 10^6 in human BM is a transplantable HSC (25); thus, the scarcity of HSCs has led to the necessity to purify the HSCs from the heterogeneous bulk population to allow their handling and manipulation. The purification of HSCs requires the detection of several cell surface markers. There are several markers whose expression is gained or lost during the differentiation process. In mice, HSCs do not express lineage markers (myeloid or lymphoid), so they are called lineage negative cells. In addition, HSCs express high levels of the stem cell antigen 1 (Sca-1) and c-Kit (CD117) marker (**Figure 1**). These cells are called LSK cells (lineage negative, Sca-1 positive, c-Kit positive) (1,26,27). As in mice, the absence of all lineage markers (Lin^-) determines human HSCs (28). As previously mentioned, CD34, expressed on 0.5% to 5% of all blood cells, was the first marker found that identifies human LT-HSCs and more differentiated progenitors (29). Later on, other markers were described in LT-HSCs such as the expression of the CD90 (Thy1) (30) or the lack of expression of CD45RA and CD38 (found to be expressed in more differentiated progenitors) (31–33). Moreover, integrin CD49f (involved in cell attachment to the extracellular matrix) was shown to be expressed in LT-HSCs (34). With respect to ST-HSCs or MPP, the loss of Thy1 and CD49f expression was proposed to be a key signature through a more differentiated state (3). Consecutive studies have classified the different populations depending on the expression of alternative markers, such as the expression of aldehyde dehydrogenase (ALDH), and other surface markers, such as CD117 or CD133 (35).

3. Gene correction of hematopoietic stem cells by addition strategies and their drawbacks

Gene therapy can be achieved by the delivery of genetic material into cells affected by a disease, and it can be accomplished by the addition, substitution, or alteration of the related genes (8). These modifications can be carried out by *in vivo* direct infusion of a vector having the ectopic

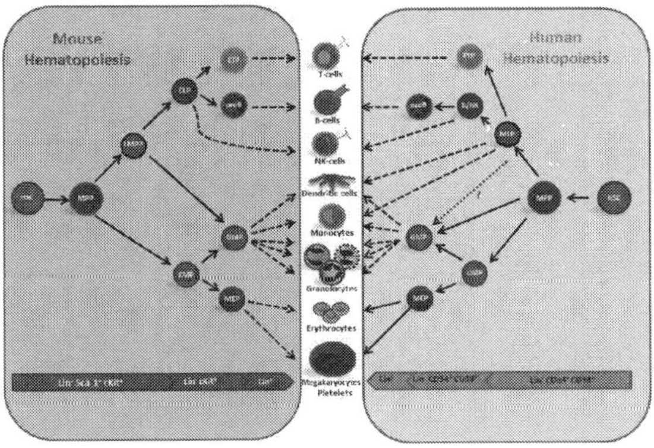

Figure 1. Human and mouse hematopoieisis. Similarities in terms of antigen differentiation markers.

therapeutic gene or by *ex vivo* manipulation of patient's cells and their reinfusion into the patient. Gene therapy has become a feasible and attractive alternative therapeutic strategy for several diseases, from inherited hematopoietic, immune, and nervous diseases, including primary immunodeficiencies, leukodystrophies, thalassemias, hemophilias, and retinal dystrophies to cancer and other malignancies (36). In the case of blood-related diseases, HSCs have long been the preferred target for *ex vivo* gene therapy due to the feasibility to isolate them *ex vivo*, genetically correct them and reinfuse them through intravenous infusion (37,38). The correction of HSCs is a promising therapeutic approach that can provide a steady and stable expression of the transgene, restoring the function of the malfunctioning gene in the patient. Stable expression of the transgene can be obtained by addition gene therapy using vector-mediated transgene insertion (**Table 1**).

Disease	Vector	Patients	Conditioning regime	Target cells	Transduced cells	Outcome	Adverse effects	References
ADA	γ-RV	Italy (18), USA (16)	Busulfan	Lymphoid cells and NK cells	HSCs	Transient or no clinical benefit. Three patients with clonal expansion with long-term efficacy followed by inactivation of the transgene	No	(41,151–153)
		UK (8)	Mephalan or busulfan					
	SIN-LV	UK (10), USA (10)	Busulfan			Unpublished, ongoing	No	–

Disease	Vector	Patients	Conditioning regime	Target cells	Transduced cells	Outcome	Adverse effects	References
SCID-X1	γ-RV	UK (11), France (11), USA (3)	None	Lymphoid cells and NK cells	HSCs	Significant T-reconstitution/T-ALL in 5 patients, 1 dead	Yes	(40,42,47,154–157)
	SIN γ-RV	UK, France, USA (9)	None			Reconstitution of functional T cells but at low number	No	(158–160)
	LV	USA	Busulfan			Ongoing		–
CGD	γ-RV	USA (10), and (3)	None or Busulfan	Neutrophils	HSCs	No benefits	Yes	(43,161–168)
		Germany (2)/ Switzerland (2)	Busulfan			Long-term correction 4/4		
		UK (4)	Melphalan			Transient correction		
		Korea (2)	Busulfan +fludarabine			Short-term correction		
	SIN γ-RV	Germany	Busulfan			Ongoing		–
	LV	UK, Switzerland, Germany, France, USA	Busulfan			Ongoing		–
SCD	SIN LV	USA	Busulfan	Ongoing	–			
MLD	SIN LV	Italy (9)	Busulfan	Macrophages and microglia	HSCs	No symptoms or just milder, after the age of symptom onset. Ongoing	No	(86)
WAS	γ-RV	Germany (10)	Busulfan	Lymphoid cells, NK cells, and platelets	HSCs	9/10 with benefits	Yes	(169,170)
	SIN LV	UK, USA, France, Italy (4)	Busulfan +fludarabine			4 with benefits, ongoing	No	(48,171,172)
β-Thalassemia	SIN LV	USA, France (4)	Busulfan	Erythroid cells	HSCs	Therapeutic benefit. Ongoing	No	(173)

Disease	Vector	Patients	Conditioning regime	Target cells	Transduced cells	Outcome	Adverse effects	References
		USA	Busulfan			Ongoing		
FA	SIN LV	USA	None			Ongoing		–
		Spain	None*			Ongoing		–
X-ALD	SIN LV	London, France	Busulfan and cyclophosphamide	Macrophages and microglia	HSCs	Therapeutic benefit.	No	(174)
		USA, France (4)				Ongoing		

Table 1. Addition gene therapy attempts and their outcome.

Classically, viral vectors, based on γ-retroviruses (γ-RV) or lentiviruses (LV), have been widely used for this purpose because they ensure high transduction efficiency and long-term expression of the transgene. γ-RVs were widely used in the first gene therapy trials that started at the end of the 1990s (39). In diseases affecting the hematopoietic system, these vectors were first used for the treatment of two inmunodeficiencies: adenosine deaminase deficiency (ADA) and X-linked severe immunodeficiency or γ-chain immunodeficiency (X-SCID1) (40–42). In the case of ADA, γ-RV-transduced autologous stem cells have been used to treat 38 patients in three independent studies (39). Overall survival resulted in 100% at 3.5 years follow-up, and from these, 20 patients no longer required enzyme replacement therapy. *Ex vivo* gene transfer in ADA patients, by means of γ-RV, demonstrated to be efficient, as the engraftment of corrected cells was maintained over time and resulted in the improvement of the cellular and humoral immune response despite the mild conditioning applied (39). Similarly, the treatment of X-SCID1 was initiated in two different clinical trials. Similar to ADA gene therapy trials, the gene complementation of γ-chain with γ-RV was clearly efficient (42). Seventeen of the 20 treated patients showed a full or nearly full correction of the T-cell defect (39). However, whereas, in ADA patients treated by gene therapy with γ-RV, no adverse effects related with the genotoxicity of the procedure were observed, in X-SCID1 patients, 5 of the 20 treated patients suffered T-cell leukemia after 2.5 to 5 years after the treatment due to insertional mutagenesis (39). Viral insertion produces insertional mutagenesis and can induce the activation of neighboring genes, as proto-oncogenes, through enhancer/promoter sequences present in the retroviral long terminal repeats (LTRs). Both γ-RV and LV contain LTRs that are present at both the 5′ and 3′ ends of retroviral RNA genome and are required for the integration of the provirus into the host genome. Similar issues arose sometimes after the treatment of other immunodeficiencies as the Wiscott-Aldrich syndrome, in which restored expression of the WASP gene was achieved, but 4 of 10 treated patients were reported to suffer from leukemia or chronic granulomatosis disease (CGD) in which the appearance of myelodysplastic syndrome in 3 patients also proved to be the result of the up-regulation of genes such as *MECOM* and *PRDM16* in CGD patients (43–47). These severe adverse effects limited the application of gene therapy to patients. A strong effort of the scientific community was required to increase the safety characteristics of the vectors. The inactivation of the U3 region in the 3′ end of the viral genome, leading to self-inactivating (SIN-LTRs) vectors (48), and the use of more physiological and/or specific promoters (49) were characteristics implemented in the new improved integrative vectors. Moreover, the analysis of insertion patterns demonstrated that γ-RV preferentially integrate near the transcriptional start sites and promoter regions (50), whereas this in LV occurs along transcriptionally active genes (51–53). Even so, some studies have reported that LV produce aberrant spliced transcript and deregulated expression in the integrated genes, although this phenomenon has not been reported in patients (54–56). Therefore, the SIN LTRs and the integration-site preferences of LV have been shown to substantially mitigate the insertional genotoxicity, making LV a reliable therapeutic option. In fact, a significant number of phase I/II clinical trials are currently being conducted with LV, which have reported remarkable efficacy and safety in immunodeficiencies, β-thalassemia, adrenoleukodystrophy, or metachromatic leukodystrophy (36). All these HSC-based gene therapy clinical trials have reported stable and high level of reconstitution of the hematopoiesis of the treated patients. Besides, there has been no report of adverse events related with LV, although the overall follow-up of some of these trials is still limited.

Nonetheless, the potential risks already observed with γ-RV could be still present. Future trials should overcome the efficient delivery of the new genetic information in the target cells, without altering or disrupting the host cell genome and preserving the characteristics of the transduced cells. The uses of LV that do not require integration or the precise control of the site where the genetic material is going to be inserted in the host genome (gene editing) are strategies that are nowadays being explored in deep. Important efforts are invested in this second approach that will overcome the majority of the potential adverse effects described above. However, additional challenges need to be overcome yet.

4. Gene editing strategies in HSCs

Gene editing describes the new technology able to accurately modify genes, either by knocking out these genes or by inserting or substituting specific sequences in a precise way. This new technology is based on promoting the action of endogenous DNA repair pathways at the target sequence without altering any other place in the genome. The main approach used for gene targeting is based on the natural HR DNA repair mechanism of the cell. Homologous Recombination (HR) is an accurate DNA repair mechanism that uses the sister chromatic as a template to repair double-stranded breaks (DSBs). HR-based gene therapy strategies started in 1980s using DNA donors flanked by long arms with homology to the target locus (57). However, the probability of inducing the specific insertion in the target site was low, around 10^{-6} (58), making this procedure only feasible for basic research. Trying to improve this low efficiency, several strategies have been addressed. The most important one is the development of engineered nucleases able to induce DSBs in the specific target site, with the consequent activation of the HR, and the gene edition in a precise place of the genome.

Although transfer of designed nucleases together with a donor template can increase the efficiency, HR is not as frequent as other DSB repair mechanisms, such as non-homologous end-joining (NHEJ). This is an error-prone mechanism that binds the two DNA ends originated in the DSB, without the use of any template. Insertions and/or deletions (INDELS) can be introduced in the target site. HR process is more frequent during the S and G_2 phases of the cell cycle, so it takes place more frequently when cells are proliferating, and its permissiveness varies between cell types (59).

There are different applications based on these two DNA repair mechanisms. If NHEJ takes place, the generated INDELs could have a particular application for the stable disruption of specific target genes (59). When HR takes place, three strategies of gene editing can be envisaged:

- Gene correction: One or more bases from the mutated original strand are replaced by the integration of the corrected ones. This strategy is the one to be chosen when the objective is to introduce or repair point or small mutations. If the mutations causing the disease are not recurrent or are produced in different genes, it will be restricted to a limited number of patients.
- Knock-in: A partial cDNA of the gene of interest is introduced in its former locus. Usually, a splicing acceptor is located before the cDNA to anchor the endogenous splicing donor. Endogenous elements of the locus will control the expression of the endogenous/exogenous chimeric cDNA/gene. Necessary to maintain the endogenous regulation of transgene.
- Safe harbor integration: It consists on the insertion of a whole expression cassette (promoter, transgene and regulatory signals) in a concrete place of the genome. This approach could be used for mutations produced in the same gene and also for mutations produced in different genes.

5. Gene editing tools applicable to the gene editing of hematopoietic stem cells

Different systems have been developed to lead the genetic modification to a specific locus. These gene editing tools have been inspired by natural systems, which evolved over millions years to modify DNA, among which the capability of generating DSBs has been selected for different purposes including DNA repair or cell response against foreign DNA from pathogens. The proteins involved in DSBs act by the specific hydrolysis of the DNA and are known as nucleases. These nucleases have been described in all branches and life species. As long as these proteins have been described, they have been rapidly applied as tools for gene editing. Four different types of nucleases have been more widely applied to generate DSBs: homing endonucleases (also called meganucleases), zinc-finger nucleases (ZFNs), transcription activator-like effector nucleases (TALENs), and clustered regularly interspaced short palindromic repeats (CRISPR)/Cas9.

5.1. Meganucleases

Meganucleases tend to be small proteins (<40 kDa) encoded by mobile inteins or introns. These families of proteins have evolved to target sequences (14–40 bp) in a specific way (60,61). In nature, meganucleases propagate their own introns cleaving cognate intronless alleles and therefore producing HR. For that capability of moving its genetic sequence in a determined place, meganucleases are also called homing endonucleases. There have been described five families of meganucleases: GIY-YIG His-Cys box, HNH, LAGLIDADG, and PD-(D/E) XK. The largest family is known as LAGLIDADG, named after the consensus sequence of its catalytic motif LAGLIDADG. It is a diverse group of enzymes present in fungal and protozoan mitochondria, plants and algae chloroplasts, bacteria, and *Archaea*. This family is defined for having either one or two copies of LAGLIDADG motif acting therefore as separated monomers or homodimers, respectively. However, meganucleases are limited as gene editing tools due to the necessity of finding large palindromic DNA targets (62). Moreover, its design requires important bioinformatics, difficult protein engineering, and *in vitro* testing.

5.2. ZFNs

In 1996, Chandrasegaran et al. started to use hybrid nucleases which combined the cleavage domain of *Fok*I nuclease and the specificity of Cys_2His_2 zinc-finger protein domains, making possible the specific DNA cut in a sequence of more than 18 bp (63,64). *Fok*I is capable of generating DSB when dimerized, so to use it as a gene editing tool, two monomers must be designed and delivered separately. Designing the two monomers in inverted orientation improved the nuclease efficiency (65). The specific domain that binds to the DNA is formed by tandem repetitions of Cys_2His_2, which form the $\beta\beta\alpha$ structure folded for the presence of a Zn^{2+} ion. The α-helix recognizes a specific DNA triplet. Between 3 and 6 of these motifs, which are separated by some spacer nucleotides, are modularly assembled in each monomer. Between the recognition sites of both strains remains a small nontargeted region called "spacer" surrounded by the *Fok*I dimer (66–68). To avoid some off-target events already described (69,70), monomers of two different *Fok*I domains, only active when forming heterodimers, have been designed (71,72). Although different ZFN open-source libraries have been developed, the design still results complicated for nonspecialists (73,74). However, ZFNs have been the first described nucleases and are the only ones already in clinical trials (https://goo.gl/M0ZWoB).

5.3. TALEN

TALEN technology follows the same rationale as ZFN, the use of *Fok*I domain because of its nonspecific cleavage capability fused to a domain that confers DNA-binding specificity. As in the case of ZFN, the *Fok*I region requires dimerization to be active, so this tool also requires the use of two monomers (65,75). In TALEN, the specificity is provided for protein domains from TALE. TALE are bacterial proteins, first described in *Xanthomonas,* capable to modify the gene transcription of the plants that these bacteria parasite (76). This DNA-binding domain is formed by 10 to 30 repetitions of 30 residues. The 12th and 13th positions, the so-called repeat-variable di-residues (RVDs), are the polymorphism responsible of the specificity (77). Each RVD pair is able to recognize specifically a base pair (77–80). Knowing the code TALE-to-DNA (79), the designing has become a relatively easy task (81). Moreover, due to the long DNA sequences they recognize, they show the lowest off-target cleavage. On the contrary, their highly repetitive structures make these sequences hard to package into viral vectors because of their tendency to recombine (82).

5.4. CRISPR/Cas9

The CRISPR/Cas9 system is the latest nuclease that has appeared in the gene editing tool arena. CRISPR/Cas9 is part of the adaptive immune system of bacteria and *Archaea*. In nature, this system cleaves foreign DNA to avoid cellular threats such as viral attacks (83,84). Up to 10 CRISPR/Cas9 systems have been described, but the type II CRISPR/Cas9 is the most used in gene engineering to generate DSBs for research.

This strategy is based in three elements: the Cas9 nuclease, the trans-activating CRISPR RNA (tracrRNA), and the CRISPR RNA (crRNA). Cas9 is a nuclease that produces DSBs, thanks to its two catalytic regions, Ruv and HNH. Cas9 recognizes the tertiary RNA structures that the duplex tracrRNA and crRNA form when they are bound to the DNA. The tracrRNA is a fix structure that binds Cas9, the catalytic element, to the crRNA, the element that gives the specificity to the target site. The crRNA is a 20-bp-long RNA strand complementary to the target region. The only restriction of the system is that the target sequences must be followed downstream for the 3-bp-long protospacer adjacent motif (PAMs), 5′-NGG-3′ (72). PAM motif is frequent along the DNA, so CRISPR/Cas9 system crRNA can be design against almost every DNA region. To facilitate its use, crRNA and tracrRNA have been combined in one single molecule called single guide RNA (gRNA) (85). Thus, only two elements must be delivered into the cells, the Cas9 and the gRNA, instead of three. Although controversial, some studies have reported a higher off-target activity than other nucleases (86). CRISPR/Cas9 system is developing fast and new Cas proteins recognizing different PAM sequences, having smaller protein sizes or even been able to cut in an overhang way have been described over the last year (87–89). Moreover, Cas9 protein modifications, as ZFN and TALE, have allowed not only the generation of DSB but also as function as DNA activators, DNA repressors, or DNA tracers (90).

6. Delivery of gene editing tools to HSCs: the main challenge

One of the most important steps to consider when performing a gene editing strategy is the delivery of the gene editing tools. In fact, nowadays this is probably the main bottleneck in gene therapy in general and definitely in the gene editing of HSCs (91).

The main objective of delivering the nucleases effectively in cells is to promote a hit-and-run activity of the enzyme that allows a short-term boost of expression that leads to a specific generation of a DSB in a short period of time to avoid toxicity and off-target activity due to the long-term expression of the nuclease. In case of performing a gene addition/gene correction strategy, we should also consider the delivery of the DNA molecule carrying the desired sequences to be integrated in the cell genome, the so-called donor DNA molecule. Taking all these facts into account, the delivery of gene editing tools can be classified in two different groups: viral and nonviral transfection methods (**Figure 2**).

Concerning the nonviral transfection methods, the gene editing tools can be delivered as plasmid DNA, *in vitro* transcribed mRNA, or purified proteins. The delivery of plasmid DNA is a widely used standard technique (92,93), as this molecule is easily engineered and easy to handle. Both gene editing tools and donor sequences can be transfected in this way. The donor sequence can be a linear donor sequence with less than 50 bp of homology (94), as well as single-stranded DNA oligonucleotides (95), especially when inducing mutations or corrections at the target site. Due to the negative charge of both plasmid DNA and the cell surface, the uptake of DNA into the cells is restricted, and the delivery needs to be supported by complexing DNA with chemicals such as calcium phosphate, coat lipids, or cationic polymers (92) or by subjecting the cells to an electric field that open pores in the nuclear and/or plasmatic membrane of the cells (electroporation). Another important factor to take into account is the target cell. In our case, the HSC is a hard-to-transfect cell. The only technique available nowadays to introduce these tools inside this cell type is nucleofection. Nucleofection is an electroporation-based transfection method that enables transfer of nucleic acids into difficult-to-transfect cells by applying a combination of electrical parameters with cell type-specific reagents. However, this process produces high cellular toxicity and even cell death depending on the plasmid size and the total amount of transfected DNA. Moreover, the delivery of plasmid DNA could mediate its random insertion into the genome of treated cells (96). The risk of random plasmid integration mainly depends on the plasmid quality, but it is also influenced by specific properties of the targeted cells, including the prevalence of DSBs and the status of their DNA repair pathways. Plasmid integration, although is a nonusual event, represents an obstacle for gene targeting approaches, as mentioned above for the more conventional retroviral vector delivery systems.

Alternative platforms to deliver nucleases have been explored, such as the transfection of *in vitro*-transcribed mRNA (97,98). The main advantage is that mRNA establishes a short-term boost of enzyme activity with low toxicity and no risk of integration (99). Thus *in vitro* transcribed mRNAs that encode ZFNs, TALENs, or Cas9 and gRNAs are the preferential forms in which these nucleases are nucleofected into HSCs. Another possibility is to deliver nucleases as proteins; this approach presents the same advantages as mRNA delivery, as they also act in a hit-and-run fashion. Genetic fusion of recombinant proteins to positively supercharged moieties favors their uptake by cellular internalization mechanisms (100,101). However, generation of high yields of soluble and active protein transduction domains (PTDs)-containing nucleases is difficult (102). An alternative strategy consists on the chemical conjugation of nucleases to PTDs for receptor-mediated endocytosis (103). Interestingly, due to the positive charge of their Cys_2-His_2 zinc-finger motifs, ZFNs display an intrinsic cell-penetrating capacity, which can lead to targeted mutagenesis (102). Other delivery options under investigation include protein transfection procedures such as electroporation (104). It has been successfully applied in the case of CRISPR/Cas9 nucleases, where HSCs have been efficiently edited by the direct nucleofection of the Cas9 protein/gRNA ribonucleoprotein complex (105). Chemical transfection agents are also being investigated for nucleases as proteins (106).

Apart from nucleofection, nonintegrating viral vectors, such as integrase-deficient LV vectors (IDLVs), adenovirus vectors (AdVs), and adeno-associated virus vectors (AAVs), also serve

http://dx.doi.org/10.5772/62383

as a robust source to facilitate the delivery of the genes encoding programmable nucleases both *in vitro* and *in vivo* in cell types that are sensitive to transfection such as HSCs (107). The decision whether to use one type or another relies on the size of the transgene to deliver and on the capability of the virus to transduce quiescent or diving cells.

AdVs are double-stranded DNA viruses capable of packaging up to 37 kb of DNA. However, the transduction capacity of these vectors is highly dependent on the cellular type. Owing to the large cargo size that can be accommodated in adenoviruses, they are an attractive platform for the delivery of programmable nucleases and donor DNA. They have been used in some gene targeting approaches to deliver nucleases in both fibroblasts and $CD4^+$ T cells (108,109), but unfortunately they are not able to efficiently transduce HSCs.

AAVs are nonenveloped, single-stranded DNA viruses that replicate only in the presence of a helper virus, such as adenoviruses or herpes simplex viruses. Nowadays, they are the most widely used delivery system for *in vivo* gene therapy due their nonpathogenic behavior and their capability to overcome the host's preexisting immunity. The major disadvantage of AAVs is their low packaging capacity that is incompatible with some therapeutic gene sizes. ZFNs are the most compact nucleases, and sequences encoding ZFNs can be packaged into these vectors (110). In the case of the CRISPR/Cas9 tool, only one of the required elements can be packaged in an AAV. The *Streptococcus pyogenes* Cas9 can be packaged into an AAV, but the gRNA sequence should go in a different vector. An alternative to the limited packaging size of the AAV is the use of the *Staphylococcus aureus* miniCas9, which perfectly fits the AAV size with the gRNA sequence. However, the proper definition of the target sites for this miniCas9 is still being improved (87).

IDLVs are currently the preferred tool for transferring genes into HSCs, as they possess certain advantageous attributes compared to other vectors [reviewed in (38)]. For example, the preintegration complex of LV is actively translocated into the nucleus and thereby facilitates efficient transduction of a variety of nondividing cells. In contrast, other viral vectors depend on the dissolution of the nuclear membrane during mitosis for delivering their cargo into the target cell nucleus. Consequently, efficient transduction of HSCs can be achieved with IDLVs after a shorter incubation time *in vitro*, preserving to some extent the physiological nature of HSCs and their engraftment potential. IDLVs have successfully delivered ZFNs and homologous donor DNA (111). Unfortunately, IDLVs are incompatible with TALENs because highly homologous TALE repeats often lead to unwanted recombination in cells (112). Moreover, the production of IDLVs is still the major drawback. The high viral titers needed to efficiently transduce HSCs are difficult to get, limiting their potential use in many instances.

7. Evidences of gene editing in HSCs and future attempts

Since gene editing technology has started to be used, its applicability to HSCs is being explored. Either blood disease modeling or new gene therapy approaches have risen from the basics of gene edition. The development of genome editing technologies based on programmable nucleases has substantially improved the ability to make precise changes in the genomes of the cells compared to small-molecule therapies. However, the gene editing advancement in hematopoiesis could not be understood without its combination with cell reprogramming technology, especially iPSCs. The correction by gene editing of patient-specific iPSC has facilitated the spreading of this technology, allowing the optimization of tools and techniques on this difficult but easily expandable cell type, which have been then applied to the more difficult to culture and transfect/transduce HSCs. On the contrary, the current clinical trials addressed to correct monogenic blood disease by viral vectors, retro and LV, have contributed to the useful knowledge about these diseases, different delivery systems, and their therapeutic potentials (Table 2).

The first evidence of gene editing in human hematopoietic progenitors was described in 2002 by Dr. Davis's group in sickle cell disease (SCD) HSCs (113). SCD is characterized by a single-point mutation in the seventh codon of the β-globin gene, which means that the site-specific correction of the mutation would allow for the production of normal hematopoietic cells. By DNA microinjection, chimeric oligonucleotides designed to direct a site-specific nucleotide exchange in the human β-globin gene were introduced into $CD34^+$ and Lin^- $CD38^-$ cells (purified from umbilical cord blood from healthy donors). Conversion rates of 10% to 15% were confirmed in Lin^- $CD38^-$ cells. These levels of correction are enough to correct the phenotype of the patients in this case, as 10% normal hematopoietic cells are sufficient to render patients free of SCD. However, human engraftment of edited HSCs in immunodeficient animals was not tested. The efficacy of gene editing in $CD34^+$ was later improved by the use of sequence-specific endonucleases. Dr. Naldini's group was able to identify gene edition in up to 80% hematopoietic progenitors after transducing cord blood $CD34^+$ with an IDLV carrying a puromycin-resistant gene together with ZFN targeting CCR5 locus and selecting them with puromycin (114). The first therapeutic application of gene editing was described 7 years later by the same group. Human HSCs were gene edited at the *IL2RG* locus, which is mutated in SCID-X1 patients, by an IDLV delivering a donor template to perform a knock-in strategy in combination with the electroporation of ZFN mRNAs against the same locus (115). Since the donor templates carried the GFP reporter cassette alone or together with the therapeutic *IL2RG* cDNA, the targeted human hematopoietic could be followed by GFP expression. GFP could be detected in up to 95% of immunodeficient mice transplanted with the human progenitors after their gene editing at short term after transplantation. However, the most significant achievement was the detection of the targeted human hematopoietic cells in 46% of the transplanted mice at long-term, which meant the targeting of the most primitive HSCs, the LT-HSCs, and opens up a real potential clinical use of gene editing in HSCs. More importantly, the IL2RG functionality was restored in T and NK cells derived from the hematopoietic progenitors edited at *IL2RG* locus. This is the first demonstration of the potential clinical use of gene edition to treat monogenic blood diseases.

Although the use of large donors to correct patient's hematopoietic progenitors through a knock-in or safe harbor approaches have got good preclinical data, its overall efficacy is still low to be applied in clinical trials. Alternatively, the use of single-stranded oligonucleotides (ssODNs) for HR has been also explored to perform gene correction due to their high efficacy and low toxicity. As Proof of concept, gene correction experiments using β-globin ssODNs as donors to correct β-thalassemia were performed. Intracellular delivery was done using biodegradable nanoparticles loaded with triplex-forming peptide nucleic acids (PNAs). Although low levels of efficacy were reached (<1%), *in vivo* analysis showed engraftment of edited cells in NOD-SCID IL2rγ-null mice (116).

Recently, another approach was tested for the introduction of a mutation able to correct SCD (117). ZFNs were designed to specifically cleave at the β-globin locus with nonrelevant off-target effects and introduced in $CD34^+$ cells (isolated from umbilical cord or peripheral blood) by electroporation. Simultaneously, the delivery of an IDLV or DNA oligonucleotide homologous donor template allows high levels of gene modification (10–20%) and engraftment in immunocompromised NSG mice.

The delivery of nucleases using AAV vectors has also demonstrated their feasibility for human HSCs gene editing. Their low toxicity and relatively easy manufacture, together with the ability of achieving high recombination frequencies with small homology regions (200 bp), make them feasible genetic instruments (118). Recent studies showed that AAV serotype 6 was efficient for the delivery of donor DNAs in human $CD34^+$ Hematopoietic Progenitor Stem Cells (HPSCs) and can induce Homologous direct Repair (HDR) when used in combination with different nucleases (119). When AAV6 was used to deliver a GFP cassette flanked by arms homologous to CCR5 or AAVS1 locus, followed by electroporation with ZFN mRNA, HDR-mediated insertion at the CCR5 or AAVS1 locus was found in 17% to 26% of cells. Studies in immunodeficient mice showed that modified cells persisted in secondary transplant recipient animals, meaning that edited cells contained long-term HSCs.

Using a different gene editing approach, β-thalassemia correction has been achieved in a preclinical study without the need of a donor template (120). This clever approach was based in the knockout of the *BCL11A* gene using ZFN to generate INDELs in the mentioned gene. This gene regulates the transition from fetal globin (γ-globin) to an adult one (β-globin). Therefore, the disruption of *BCL11A* gene was able to ameliorate β-thalassemia due to increasing γ-globin and reducing mutated β-globin level.

The transfer of programmable nucleases into HSCs has been also developed as antiviral therapies by deleting genes encoding receptors essential to viral entry in the host cells, thus preventing the infection. The definitive cure of HIV or the achievement of efficient protection against this virus constitutes one of the biggest goals of modern medicine. Gene therapy can bring us closer to this purpose by the modification of *CCR5*, an HIV-1 coreceptor involved in the entrance of the viral particles in hematopoietic cells. The introduction by electroporation of specifically designed ZFNs against the *CCR5* gene attained 17% of allele disruption. This ZFN-treated HPSCs were capable of engraft in NOD-SCID IL2rγ-null mice and were able to produce a polyclonal multilineage progeny where *CCR5* disruption was maintained (121). Moreover, when mice transplanted with ZFN-modified cells were challenged with the HIV-1 virus, the animals were able to maintain the levels of $CD4^+$ T cells throughout their tissues and showed lower levels of HIV-1 viral load compared to the control animals engrafted with nonedited HPSCs. In the same direction, other strategies such as the introduction of CCR5-Specific ZFNs or CCR5-MegaTALEN mRNAs into HPSCs using adenoviral vectors and electroporation, respectively, have demonstrated the potential of this (118,122).

Genome editing in HSCs has also been used to model cancer generation and development. For example, $CD34^+$ cells purified from human cord blood were capable of initiating leukemia in response to endogenous activated *MLL-AF9* or *MLL-ENL* oncogenes, generated by TALEN-mediated knock-in (123,124). These cells displayed altered *in vitro* growth potentials and induce acute leukemias following transplantation in immunocompromised mice at a mean latency of 16 weeks. Additionally, the phenotype, morphology, and molecular features of the induced leukemias were similar to patient leukemic blasts. Consequently, this strategy provides an experimental platform for prospective studies of mixed lineage leukemia (MLL). Similarly, the CRISPR/Cas9 system is also applicable to generate some cancer models. Translocations resembling those described in acute myeloid leukemia and Ewing's sarcoma (*RUNX1/ETO* and *EWS/FLI*, respectively) were generated in human cell lines and primary cells. Cas9 nuclease and the specific gRNA were introduced as plasmidic DNA by electroporation, reaching 4% efficiency in the generation of cancer-like translocations (125).

The application of gene editing to correct or study blood diseases is a growing field, where a rising number of publications exploring all the potential of this precise gene modification technology are being generated rapidly. We should be ready for the burst of gene editing in hematopoiesis.

8. Gene-edited HSCs goes to the clinic

A notable number of phase I/II gene therapy clinical trials have reported evidence of efficacy and safety for the treatment of various genetic diseases of the blood, including primary immunodeficiencies, leukodystrophies, thalassemia, and hemophilia. However, some difficulties are acting as a bottleneck in the clinical approach of gene editing tools. First, the delivery of genetic tools into hematopoietic progenitors must be improved. Second, an advance in the culture conditions is needed, given that gene-corrected cells must be present in large enough quantities to revert the condition and transmit the modification to their progeny. In this way, some compounds capable of preserving the "stemness" of $CD34^+$ HPSCs are nowadays being explored, such as SR-1 or UM171, with promising results (126).

Up to now, only one clinical trial worldwide is open to apply gene editing in human HSCs. The trial is addressing the safety of the transplantation of hematopoietic progenitors, gene edited for the *CCR5* disruption using specific ZFNs, into HIV-1-infected patients. These patients have undetectable virus but suboptimal $CD4^+$ cell levels. Phase 1 trial is ongoing and will provide first in human data about the use of gene editing tools (https://goo.gl/M0ZWoB).

9. Other alternative nonhematopoietic sources of HSCs and their edition

Gene edited autologous HSC transplantation (HSCT) approaches remain elusive because of the difficulty in the culture and expansion of long-term repopulating HSCs *ex vivo*, as these protocols still extend for long periods of time to achieve the delivery of the correcting tools and the gene edition itself. An alternative therapeutic strategy is the use of patient-derived induced pluripotent stem cells (iPSCs) in which disease-causing mutations are corrected by gene targeting. The development of iPSC-based approach provides an unlimited source of subject-derived cells that bypasses the main limitation of HSCs-based protocols, as iPSCs possess properties of self-renewal and pluripotency that are similar to those of embryonic stem cells (127–129). Moreover, the availability of HSCs is compromised in some situations, and in this context, easily accessible cell sources, such as skin fibroblasts (130), keratinocytes (131), or peripheral blood mononuclear cells (132), constitute an advantage for gene editing protocols. These corrected iPSCs could then be differentiated into the desirable hematopoietic cell for transplantation into patients to treat the disease (133). Here are some of the most relevant attempts following this strategy.

Regarding erythroid lineage defects, efforts have focused on β-globin alterations causing β-thalassemia and SCD. Most of the works have reprogrammed patient-derived cells by transduction of reprogramming factors, in the case of skin fibroblasts (134–137), or by nucleofection of amniotic fluid-derived cells (138,139). The gene editing strategy selected was a knock-in approach, and in the case of β-thalassemia patient-derived iPSCs, the delivery of TALENs (138,139) or CRISPR/Cas9 (135,139,140) by electroporation achieved specific gene targeting in human hemoglobin subunit β (*HBB*) locus, with variable HR rates. More precisely, in a direct comparison of CRISPR/Cas9 and TALEN, the first ones induced DSBs with greater efficiency than TALEN, but the second ones mediated a higher homologous gene targeting efficiency (139). In all the cases, the expression of *HBB* in colonies formed after clonogenic progenitor cell assays was comparable to endogenous β-globin allele, but the main challenge was the complete erythroid differentiation of the edited iPSCs, as embryonic and fetal but not adult hemoglobin was expressed (135,139). Singularly, Wang et al. reported that gene correction without assistant nucleases of iPSC reprogrammed from patient-derived fibroblasts and then differentiated to hematopoietic progenitors could differentiate *in vivo* and produce human β-globin in irradiated SCID mice, although the efficacy of correction was lower than 1% (134).

In the case of SCD, patient-derived fibroblasts were reprogrammed to iPSCs and subsequently edited by nucleofection of ZFN in a knock-in strategy. Recombination efficacy ranged from 1% to 10%, and as it happened in thalassemia studies, the vast majority of hemoglobin expression was of the fetal type (136,137).

The gene editing of iPSCs derived from peripheral blood mononuclear cells from pyruvate kinase deficiency (PKD) patients assisted by TALENs in a knock-in strategy rendered approximately 10% of efficacy in gene editing, obtaining high numbers of erythroid cells with a normalized ATP level and a normal metabolic function, providing an approach in which cells can undergo up to $2{\times}10^4$ -fold expansion to correct metabolic erythroid diseases (9).

In the case of primary immunodeficiencies, the access to BM and thymocyte samples from untreated patients with SCID is challenging, as these conditions are rare and infants typically presenting with life-threatening infections require urgent HSCT to survive (141). Different SCIDs have been addressed by gene editing of patient-derived iPSCs. Chang et al. demonstrated that locus-specific correction of the human JAK3 mutation by CRISPR/Cas9-enhanced gene targeting of iPSCs derived from skin keratinocytes of SCID patients restored the differentiation potential of early T-cell progenitors. The strategy used was knock-in and efficacies ranged from 6% to 73 % of the analyzed clones when using one gRNA and from 0 to 100% of G418-resistant clones when using two guides. Corrected progenitors were capable of producing NK cells and mature T-cell populations expressing a broad TCR repertoire as well as all hematopoietic lineages. However, after transplantation in NSG mice, the differentiation to all hematopoietic lineages was not detectable (141).

Human mesenchymal cells from SCID-X1 patients were reprogrammed to iPSCs and afterwards corrected by HR assisted by TALENs in the *IL-2Rg* locus, following also a knock-in strategy, with an efficiency of 2.6% (142). They used a selection-free approach, as cells bearing a functional γ-chain show a positive selective advantage (143). The gene editing resulted in the rescue of T-cell precursors and mature NK cells after long-term differentiation *in vitro* (142). In ADA-SCID patients, the protocol was improved to perform reprogramming and gene targeting together in a one-step procedure that required only a single electroporation in an attempt to avoid cells to be in culture for several months, which is not compatible for patients for whom urgent medical intervention is imperative. They obtained 5% corrected iPSCs from skin biopsies of ADA-SCID patients after gene editing assisted by CRISPR/Cas9 and using single-stranded corrective oligonucleotide (ssODN) as therapeutic matrix, where the expression of corrected ADA mRNA was confirmed (144).

Some works have been performed in X-linked chronic granulomatous disease (X-CGD), being the selected strategy for gene editing the directed insertion of therapeutic cDNA in the safe harbor locus *AAVS1*. The selection of the promoter that drives the expression of the therapeutic gene seems to be quite important in CGD, as an inappropriate expression of the transgene in other cellular types, as the stem cell compartment, could lead to an exacerbate reactive oxygen species (ROS) production and engraftment failure (43,145). Merling et al. have developed a platform that combines the production of iPSCs derived from patients with the five genetic forms of CGD with the targeting of *AAVS1* safe harbor assisted by ZFNs, resulting in neutrophils and macrophages differentiated from the corrected iPSCs with a restored ROS production and antimicrobial function (146).

TALEN-assisted specific gene correction of *RUNX1* gene in iPSCs derived from dermal fibroblasts from a patient with familial platelet disorder resulted in the restoration of megakaryopoiesis and subsequent maturation of the corrected cells (147). Even in a DNA repair deficiency syndrome such as Fanconi anemia, with defects in homology-directed DNA repair, the wild-type cDNA of *FANCA* was targeted to *AAVS1*, assisted by ZFNs with an efficiency of gene targeting up to 4%, rendering disease-free hematopoietic progenitors (108).

Direct reprogramming, meaning the generation of hematopoietic cells or HSCs directly from nonhematopoietic cells, has been also assayed with only modest or even irreproducible results (148,149). No data are available related to the combination of gene correction with hematopoietic direct reprogramming.

Taking all these data into account, an iPSC-based approach would provide an unlimited source of patient-derived corrected cells from which hematopoietic cells could be derived continu-

http://dx.doi.org/10.5772/62383

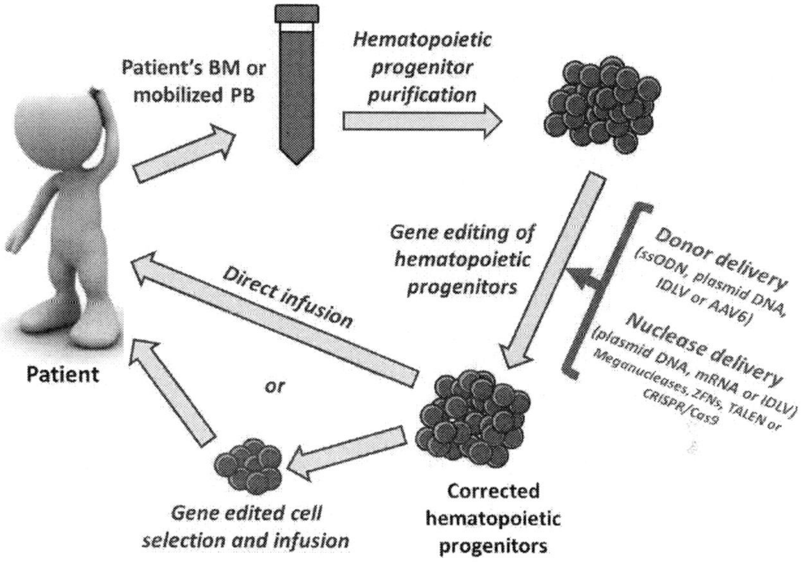

Figure 2. Potential strategies to follow in a HSC gene editing clinical protocol.

ously, and this approach could serve as a complementary approach to treat patients, for example, infusions of *in vitro*-derived autologous T cells to stabilize patients after HSCT (150). However, proper differentiations of the desirable cellular type, being the generation of real HSCs the major barrier, and an unsuitable engraftment of corrected cells remain the main pitfalls for the translation of this technology to the clinics.

10. Overall remarks

Although conventional gene therapy is becoming an alternative therapeutic option, being considered in some instances before allogeneic BM transplantation therapeutic option, the development of new tools and strategies to overcome the potential risks associated with the present ones is an already established challenge in the gene therapy field. Moreover, the precise correction of the mutation or mutations present in the patient's cells is the primary objective of any researcher committed to gene therapy. In this respect, gene editing is clearly an option to be explored and optimized. The gene editing field is constantly growing and changing. In the future, both hereditary and acquired diseases may benefit from gene editing strategies coupled with hematopoietic cell transplantation. As the conventional gene therapy of monogenic disease is the first and most successful clinical therapy involving genetic modification, the partnership between gene editing and hematopoietic progenitor transplantation will be one of the widest and effective gene therapies for these types of hematopoietic diseases.

Without any doubts, this traces the origin of a new era in cell and gene therapy and brings us closer to the aim of accurate site-specific gene editing in patients.

Cell source	Nuclease	Donor	Delivery system (nuclease/ donor)	Strategy	Efficiency	Gene therapy	Gene edited HSC engraftment	Reference
Human UCB-CD34⁺ and Lin⁻ CD38⁻	No	Chimeric oligonucleotides	Microinjection	Gene correction	10–15%	Modeling point mutation of sickle-cell anemia	Not tested	(113)
Human UCB-CD34⁺	ZFN	CCR5 HA-PGK-GFP-CCR5 HA	IDLV/IDLV	Safe harbor (CCR5)	0.03–0.11%	Proof-of-concept	Not tested	(114)
Human mobilized PB-CD34⁺	No	β-Globin ssODN	PNAs	Gene correction	0.5–1%	Proof-of-concept	Not tested	(175)
Human mobilized PB-CD34⁺	No	CCR5 and β-globin ssODN	PNAs	Gene correction	0.43%	Proof-of- concept	Yes (in vivo gene editing of human hematopoietic cells engrafted in NSG)	(116)
Human UCB-CD34⁺ and BM-CD34⁺	ZFN	HA-PGK-GFP-HA and HA-IL2RG-cDNA-PGK-GFP-HA	mRNA electroporation/ IDLV	(AAVS1) and knock-in	3–11%	SCID-X1 patient's HSC	Yes (NSG)	(115)
Human mobilized PB-CD34⁺	ZFN	No	mRNA electroporation	Knockout (BCL11A)	60–80%	β-Thalassemia	Not tested	(120)
Human BM-CD34⁺	ZFN	β-Globin donor IDLV or ssODN	mRNA electroporation/ IDLV or ssODN electroporation	Gene correction	30–40%	Sickle-cell anemia	Yes (NSG)	(117)
Human mobilized PB-CD34⁺	ZFN	HA-GFP-HA or ssODN donor AAV6	mRNA electroporation/	Safe harbor (CCR5 and AAVS1)	17–26% or 19–43%	Proof-of-concept	Yes (NSG)	(119)

Cell source	Nuclease	Donor	Delivery system (nuclease/ donor)	Strategy	Efficiency	Gene therapy	Gene edited HSC engraftment	Reference
or fetal liver CD34+			AAV6 donor delivery					
Human mobilized PB-CD34+	ZFN	No	Plasmidic DNA electroporation	Knockout (CCR5)	17%	HIV protection	Yes (NSG)	(121)
Human mobilized PB-CD34+	ZFN	No	Adenoviral vector	Knockout (CCR5)	26–31%	HIV protection	Yes (NSG)	(176)
Human mobilized PB-CD34+	CRISPR/ Cas9	No	Plasmidic DNA electroporation	Knockout (B2M and CCR5)	26.8%	Hypoimmunogenic cells for transplantation and HIV protection	Yes (NSG)	(177)
Human mobilized PB-CD34+	ZFN/ TALEN	No	Adenoviral vector	Knockout (CCR5 and HS2-globin locus control region)	8.4–12%	HIV protection	Yes (NSG)	(122)
Human mobilized PB-CD34+	megaTAL	CCR5 HA-MND-GFP-CCR5 HA	mRNA electroporation/ AAV	Knockout (CCR5)	15–20%	HIV protection	Yes (NSG)	(178)
Human BM-CD34+	ZFN	No	Permeability mediated by multinuclear localization signal (multi-NLS ZFN)	Knockout (CCR5)	17%	HIV protection	Not tested	(179)
Human UCB-CD34+	CRISPR/ Cas9	No	Plasmidic DNA electroporation	Chromosomal translocation (RUNX1/ ETO)	0.96–4.07%	Cancer modeling	Not tested	(180)
Human UCB-CD34+	TALEN	MLL HA-cDNA(AF9, ENL2 or ENL7)-IRES-NeonGreen-MLL HA	Plasmidic DNA electroporation	Chromosomal translocation (MLL-AF9, MLL-ENL2, and MLL-ENL7)	9%	Cancer modeling	Yes (NSG)	(123)

Table 2. Present results in the gene editing of hematopoietic stem cells.

Author details

Sergio López-Manzaneda, Sara Fañanas-Baquero, Virginia Nieto-Romero, Francisco-Jose Roman-Rodríguez, Maria Fernandez-Garcia, Maria J. Pino-Barrio, Fatima Rodriguez-Fornes, Begoña Diez-Cabezas, Maria Garcia-Bravo, Susana Navarro, Oscar Quintana-Bustamante and Jose C. Segovia

Hematopoietic Innovative Therapies Division, Centro de Investigaciones Energéticas, Medioambientales y Tecnológicas (CIEMAT) – Centro de Investigación Biomédica en Red de Enfermedades Raras (CIBERER). Madrid, Spain

Advanced Therapies Unit, Instituto de Investigación Sanitaria-Fundación Jiménez Díaz (IIS-FJD, UAM), Madrid, Spain

References

[1] Morrison SJ, Uchida N, Weissman IL. The biology of hematopoietic stem cells. Annu Rev Cell Dev Biol. 1995;11:35–71.

[2] Till JE, Mc CE. A direct measurement of the radiation sensitivity of normal mouse bone marrow cells. Radiat Res. 1961;14:213–22.

[3] Doulatov S, Notta F, Laurenti E, Dick JE. Hematopoiesis: a human perspective. Cell Stem Cell. 2012;10:120–36.

[4] Stolfi JL, Pai CS, Murphy WJ. Preclinical modeling of hematopoietic stem cell transplantation: advantages and limitations. FEBS J. 2015.

[5] Rogers S, Pfuderer P. Use of viruses as carriers of added genetic information. Nature. 1968;219:749–51.

[6] Bersenev A, Levine BL. Convergence of gene and cell therapy. Regen Med. 2012;7:50–6.

[7] Cavazzana M. Hematopoietic stem cell gene therapy: progress on the clinical front. Hum Gene Ther. 2014;25:165–70.

[8] Kaufmann KB, Buning H, Galy A, Schambach A, Grez M. Gene therapy on the move. EMBO Mol Med. 2013;5:1642–61.

[9] Garate Z, Quintana-Bustamante O, Crane AM, Olivier E, Poirot L, Galetto R, et al. Generation of a high number of healthy erythroid cells from gene-edited pyruvate

kinase deficiency patient-specific induced pluripotent stem cells. Stem Cell Rep. 2015;5:1053–66.

[10] Metcalf D. The molecular control of cell division, differentiation commitment and maturation in haemopoietic cells. Nature. 1989;339:27–30.

[11] Pike BL, Robinson WA. Human bone marrow colony growth in agar-gel. J Cell Physiol. 1970;76:77–84.

[12] Moore MA, Williams N, Metcalf D. *In vitro* colony formation by normal and leukemic human hematopoietic cells: characterization of the colony-forming cells. J Natl Cancer Inst. 1973;50:603–23.

[13] Gartner S, Kaplan HS. Long-term culture of human bone marrow cells. Proc Natl Acad Sci U S A. 1980;77:4756–9.

[14] Sutherland HJ, Eaves CJ, Eaves AC, Dragowska W, Lansdorp PM. Characterization and partial purification of human marrow cells capable of initiating long-term hematopoiesis *in vitro*. Blood. 1989;74:1563–70.

[15] Guezguez B, Bhatia M. Transplantation of human hematopoietic repopulating cells: mechanisms of regeneration and differentiation using human-mouse xenografts. Curr Opin Organ Transplant. 2008;13:44–52.

[16] Bosma GC, Custer RP, Bosma MJ. A severe combined immunodeficiency mutation in the mouse. Nature. 1983;301:527–30.

[17] Fulop GM, Phillips RA. The SCID mutation in mice causes a general defect in DNA repair. Nature. 1990;347:479–82.

[18] Shultz LD, Schweitzer PA, Christianson SW, Gott B, Schweitzer IB, Tennent B, et al. Multiple defects in innate and adaptive immunologic function in NOD/LtSz-SCID mice. J Immunol. 1995;154:180–91.

[19] Ito M, Hiramatsu H, Kobayashi K, Suzue K, Kawahata M, Hioki K, et al. NOD/SCID/ γ(c)(null) mouse: an excellent recipient mouse model for engraftment of human cells. Blood. 2002;100:3175–82.

[20] Shultz LD, Lyons BL, Burzenski LM, Gott B, Chen X, Chaleff S, et al. Human lymphoid and myeloid cell development in NOD/LtSz-SCID IL2Rγ null mice engrafted with mobilized human hemopoietic stem cells. J Immunol. 2005;174:6477–89.

[21] Civin CI, Trischmann T, Kadan NS, Davis J, Noga S, Cohen K, et al. Highly purified CD34-positive cells reconstitute hematopoiesis. J Clin Oncol. 1996;14:2224–33.

[22] Larochelle A, Vormoor J, Hanenberg H, Wang JC, Bhatia M, Lapidot T, et al. Identification of primitive human hematopoietic cells capable of repopulating NOD/SCID mouse bone marrow: implications for gene therapy. Nat Med. 1996;2:1329–37.

[23] Bellantuono I. Haemopoietic stem cells. Int J Biochem Cell Biol. 2004;36:607–20.

[24] Brown JM, Weissman IL. Progress and prospects in hematopoietic stem cell expansion and transplantation. Exp Hematol. 2004;32:693–5.

[25] 25. Wang JC, Doedens M, Dick JE. Primitive human hematopoietic cells are enriched in cord blood compared with adult bone marrow or mobilized peripheral blood as measured by the quantitative *in vivo* SCID-repopulating cell assay. Blood. 1997;89:3919–24.

[26] Spangrude GJ, Heimfeld S, Weissman IL. Purification and characterization of mouse hematopoietic stem cells. Science. 1988;241:58–62.

[27] 27. Ikuta K, Weissman IL. Evidence that hematopoietic stem cells express mouse c-kit but do not depend on steel factor for their generation. Proc Natl Acad Sci U S A. 1992;89:1502–6.

[28] Bernstein ID, Andrews RG, Rowley S. Isolation of human hematopoietic stem cells. Blood cells. 1994;20:15–23; discussion 4.

[29] 29. Civin CI, Strauss LC, Brovall C, Fackler MJ, Schwartz JF, Shaper JH. Antigenic analysis of hematopoiesis. III. A hematopoietic progenitor cell surface antigen defined by a monoclonal antibody raised against KG-1a cells. Immunol J. 1984;133:157–65.

[30] Baum CM, Weissman IL, Tsukamoto AS, Buckle AM, Peault B. Isolation of a candidate human hematopoietic stem-cell population. Proc Natl Acad Sci U S A. 1992;89:2804–8.

[31] Lansdorp PM, Sutherland HJ, Eaves CJ. Selective expression of CD45 isoforms on functional subpopulations of $CD34^+$ hemopoietic cells from human bone marrow. J Exp Med. 1990;172:363–6.

[32] Bhatia M, Wang JC, Kapp U, Bonnet D, Dick JE. Purification of primitive human hematopoietic cells capable of repopulating immune-deficient mice. Proc Natl Acad Sci U S A. 1997;94:5320–5.

[33] Conneally E, Cashman J, Petzer A, Eaves C. Expansion *in vitro* of transplantable human cord blood stem cells demonstrated using a quantitative assay of their lympho-myeloid repopulating activity in nonobese diabetic-SCID/SCID mice. Proc Natl Acad Sci U S A. 1997;94:9836–41.

[34] Notta F, Doulatov S, Laurenti E, Poeppl A, Jurisica I, Dick JE. Isolation of single human hematopoietic stem cells capable of long-term multilineage engraftment. Science. 2010;333:218–21.

[35] Mirabelli P, Di Noto R, Lo Pardo C, Morabito P, Abate G, Gorrese M, et al. Extended flow cytometry characterization of normal bone marrow progenitor cells by simultaneous detection of aldehyde dehydrogenase and early hematopoietic antigens: implication for erythroid differentiation studies. BMC Physiol. 2008;8:13.

[36] Naldini L. Gene therapy returns to centre stage. Nature. 2015;526:351–60.

[37] Ghosh S, Thrasher AJ, Gaspar HB. Gene therapy for monogenic disorders of the bone marrow. Br J Haematol. 2015.

[38] Naldini L. *Ex vivo* gene transfer and correction for cell-based therapies. Nat Rev. 2011;12:301–15.

[39] Touzot F, Hacein-Bey-Abina S, Fischer A, Cavazzana M. Gene therapy for inherited immunodeficiency. Expert Opin Biol Ther. 2014;14:789–98.

[40] Gaspar HB, Cooray S, Gilmour KC, Parsley KL, Adams S, Howe SJ, et al. Long-term persistence of a polyclonal T cell repertoire after gene therapy for X-linked severe combined immunodeficiency. Sci Transl Med. 2011;3:97ra79.

[41] Gaspar HB, Cooray S, Gilmour KC, Parsley KL, Zhang F, Adams S, et al. Hematopoietic stem cell gene therapy for adenosine deaminase-deficient severe combined immunodeficiency leads to long-term immunological recovery and metabolic correction. Sci Transl Med. 2011;3:97ra80.

[42] Hacein-Bey-Abina S, Hauer J, Lim A, Picard C, Wang GP, Berry CC, et al. Efficacy of gene therapy for X-linked severe combined immunodeficiency. N Engl J Med. 2010;363:355–64.

[43] Grez M, Reichenbach J, Schwable J, Seger R, Dinauer MC, Thrasher AJ. Gene therapy of chronic granulomatous disease: the engraftment dilemma. Mol Ther. 2011;19:28–35.

[44] Modlich U, Navarro S, Zychlinski D, Maetzig T, Knoess S, Brugman MH, et al. Insertional transformation of hematopoietic cells by self-inactivating lentiviral and gammaretroviral vectors. Mol Ther. 2009;17:1919–28.

[45] Montini E, Cesana D, Schmidt M, Sanvito F, Bartholomae CC, Ranzani M, et al. The genotoxic potential of retroviral vectors is strongly modulated by vector design and integration site selection in a mouse model of HSC gene therapy. J Clin Invest. 2009;119:964–75.

[46] Hacein-Bey-Abina S, Von Kalle C, Schmidt M, McCormack MP, Wulffraat N, Leboulch P, et al. LMO2-associated clonal T cell proliferation in two patients after gene therapy for SCID-X1. Science. 2003;302:415–9.

[47] Howe SJ, Mansour MR, Schwarzwaelder K, Bartholomae C, Hubank M, Kempski H, et al. Insertional mutagenesis combined with acquired somatic mutations causes leukemogenesis following gene therapy of SCID-X1 patients. J Clin Invest. 2008;118:3143–50.

[48] Hacein-Bey Abina S, Gaspar HB, Blondeau J, Caccavelli L, Charrier S, Buckland K, et al. Outcomes following gene therapy in patients with severe Wiskott-Aldrich syndrome. JAMA. 2015;313:1550–63.

[49] Amendola M, Venneri MA, Biffi A, Vigna E, Naldini L. Coordinate dual-gene transgenesis by lentiviral vectors carrying synthetic bidirectional promoters. Nat Biotechnol. 2005;23:108–16.

[50] Wu X, Li Y, Crise B, Burgess SM. Transcription start regions in the human genome are favored targets for MLV integration. Science. 2003;300:1749–51.

[51] Schroder AR, Shinn P, Chen H, Berry C, Ecker JR, Bushman F. HIV-1 integration in the human genome favors active genes and local hotspots. Cell. 2002;110:521–9.

[52] Mitchell RS, Beitzel BF, Schroder AR, Shinn P, Chen H, Berry CC, et al. Retroviral DNA integration: ASLV, HIV, and MLV show distinct target site preferences. PLo S Biol. 2004;2:E234.

[53] Trono D. Gene therapy: too much splice can spoil the dish. J Clin Invest. 2012;122:1600–2.

[54] De Palma M, Montini E, Santoni de Sio FR, Benedicenti F, Gentile A, Medico E, et al. Promoter trapping reveals significant differences in integration site selection between MLV and HIV vectors in primary hematopoietic cells. Blood. 2005;105:2307–15.

[55] Cesana D, Sgualdino J, Rudilosso L, Merella S, Naldini L, Montini E. Whole transcriptome characterization of aberrant splicing events induced by lentiviral vector integrations. J Clin Invest. 2012;122:1667–76.

[56] Moiani A, Paleari Y, Sartori D, Mezzadra R, Miccio A, Cattoglio C, et al. Lentiviral vector integration in the human genome induces alternative splicing and generates aberrant transcripts. J Clin Invest. 2012;122:1653–66.

[57] Capecchi MR. Altering the genome by homologous recombination. Science. 1989;244:1288–92.

[58] Cathomen T, Joung JK. Zinc-finger nucleases: the next generation emerges. Mol Ther. 2008;16:1200–7.

[59] Ciccia A, Elledge SJ. The DNA damage response: making it safe to play with knives. Mol Cell. 2010;40:179–204.

[60] Paques F, Duchateau P. Meganucleases and DNA double-strand break-induced recombination: perspectives for gene therapy. Curr Gene Ther. 2007;7:49–66.

[61] Grizot S, Epinat JC, Thomas S, Duclert A, Rolland S, Paques F, et al. Generation of redesigned homing endonucleases comprising DNA-binding domains derived from two different scaffolds. Nucleic Acids Res. 2010;38:2006–18.

[62] Epinat JC, Arnould S, Chames P, Rochaix P, Desfontaines D, Puzin C, et al. A novel engineered meganuclease induces homologous recombination in yeast and mammalian cells. Nucleic Acids Res. 2003;31:2952–62.

[63] Kim YG, Cha J, Chandrasegaran S. Hybrid restriction enzymes: zinc finger fusions to Fok I cleavage domain. Proc Natl Acad Sci U S A. 1996;93:1156–60.

[64] Chandrasegaran S, Smith J. Chimeric restriction enzymes: what is next? Biol Chem. 1999;380:841–8.

[65] Smith J, Bibikova M, Whitby FG, Reddy AR, Chandrasegaran S, Carroll D. Requirements for double-strand cleavage by chimeric restriction enzymes with zinc finger DNA-recognition domains. Nucleic Acids Res. 2000;28:3361–9.

[66] Porteus MH, Carroll D. Gene targeting using zinc finger nucleases. Nat Biotechnol. 2005;23:967–73.

[67] Urnov FD, Rebar EJ, Holmes MC, Zhang HS, Gregory PD. Genome editing with engineered zinc finger nucleases. Nat Rev. 2010;11:636–46.

[68] Carroll D. Zinc-finger nucleases: a panoramic view. Curr Gene Ther. 2011;11:2–10.

[69] Gabriel R, Lombardo A, Arens A, Miller JC, Genovese P, Kaeppel C, et al. An unbiased genome-wide analysis of zinc-finger nuclease specificity. Nat Biotechnol. 2011;29:816–23.

[70] Pattanayak V, Ramirez CL, Joung JK, Liu DR. Revealing off-target cleavage specificities of zinc-finger nucleases by *in vitro* selection. Nat Methods. 2011;8:765–70.

[71] Miller JC, Holmes MC, Wang J, Guschin DY, Lee YL, Rupniewski I, et al. An improved zinc-finger nuclease architecture for highly specific genome editing. Nat Biotechnol. 2007;25:778–85.

[72] Jinek M, Chylinski K, Fonfara I, Hauer M, Doudna JA, Charpentier E. A programmable dual-RNA-guided DNA endonuclease in adaptive bacterial immunity. Science. 2012;337:816–21.

[73] Maeder ML, Thibodeau-Beganny S, Osiak A, Wright DA, Anthony RM, Eichtinger M, et al. Rapid "open-source" engineering of customized zinc-finger nucleases for highly efficient gene modification. Mol Cell. 2008;31:294–301.

[74] Maeder ML, Thibodeau-Beganny S, Sander JD, Voytas DF, Joung JK. Oligomerized pool engineering (OPEN): an 'open-source' protocol for making customized zinc-finger arrays. Nat Protoc. 2009;4:1471–501.

[75] Bitinaite J, Wah DA, Aggarwal AK, Schildkraut I. *Fok*I dimerization is required for DNA cleavage. Proc Natl Acad Sci U S A. 1998;95:10570–5.

[76] Boch J, Bonas U. Xanthomonas AvrBs3 family-type III effectors: discovery and function. Annu Rev Phytopathol. 2010;48:419–36.

[77] Deng D, Yan C, Pan X, Mahfouz M, Wang J, Zhu JK, et al. Structural basis for sequence-specific recognition of DNA by TAL effectors. Science. 2012;335:720–3.

[78] Bogdanove AJ, Voytas DF. TAL effectors: customizable proteins for DNA targeting. Science. 2011;333:1843–6.

[79] Boch J, Scholze H, Schornack S, Landgraf A, Hahn S, Kay S, et al. Breaking the code of DNA binding specificity of TAL-type III effectors. Science. 2009;326:1509–12.

[80] Mak AN, Bradley P, Cernadas RA, Bogdanove AJ, Stoddard BL. The crystal structure of TAL effector PthXo1 bound to its DNA target. Science. 2012;335:716–9.

[81] Mussolino C, Cathomen T. TALE nucleases: tailored genome engineering made easy. Curr Opin Biotechnol. 2012;23:644–50.

[82] Holkers M, Maggio I, Liu J, Janssen JM, Miselli F, Mussolino C, et al. Differential integrity of TALE nuclease genes following adenoviral and lentiviral vector gene transfer into human cells. Nucleic Acids Res. 2013;41:e63.

[83] Bhaya D, Davison M, Barrangou R. CRISPR-Cas systems in bacteria and Archaea: versatile small RNAs for adaptive defense and regulation. Annu Rev Genet. 2011;45:273–97.

[84] Al-Attar S, Westra ER, van der Oost J, Brouns SJ. Clustered regularly interspaced short palindromic repeats (CRISPRs): the hallmark of an ingenious antiviral defense mechanism in prokaryotes. Biol Chem. 2011;392:277–89.

[85] Cho SW, Kim S, Kim JM, Kim JS. Targeted genome engineering in human cells with the Cas9 RNA-guided endonuclease. Nat Biotechnol. 2013;31:230–2.

[86] Biffi A, Montini E, Lorioli L, Cesani M, Fumagalli F, Plati T, et al. Lentiviral hematopoietic stem cell gene therapy benefits metachromatic leukodystrophy. Science. 2013;341:1233158.

[87] Ran FA, Cong L, Yan WX, Scott DA, Gootenberg JS, Kriz AJ, et al. In vivo genome editing using *Staphylococcus aureus* Cas9. Nature. 2015;520:186–91.

[88] Kleinstiver BP, Prew MS, Tsai SQ, Topkar VV, Nguyen NT, Zheng Z, et al. Engineered CRISPR-Cas9 nucleases with altered PAM specificities. Nature. 2015;523:481–5.

[89] Zetsche B, Gootenberg JS, Abudayyeh OO, Slaymaker IM, Makarova KS, Essletzbichler P, et al. Cpf1 is a single RNA-guided endonuclease of a class 2 CRISPR-Cas system. Cell. 2015.

[90] Mali P, Esvelt KM, Church GM. Cas9 as a versatile tool for engineering biology. Nat Methods. 2013;10:957–63.

[91] Skipper KA, Mikkelsen JG. Delivering the goods for genome engineering and editing. Hum Gene Ther. 2015.

[92] Urnov FD, Miller JC, Lee YL, Beausejour CM, Rock JM, Augustus S, et al. Highly efficient endogenous human gene correction using designed zinc-finger nucleases. Nature. 2005;435:646–51.

[93] Porteus MH, Baltimore D. Chimeric nucleases stimulate gene targeting in human cells. Science. 2003;300:763.

[94] Orlando SJ, Santiago Y, DeKelver RC, Freyvert Y, Boydston EA, Moehle EA, et al. Zinc-finger nuclease-driven targeted integration into mammalian genomes using donors with limited chromosomal homology. Nucleic Acids Res. 2010;38:e152.

[95] Chen F, Pruett-Miller SM, Huang Y, Gjoka M, Duda K, Taunton J, et al. High-frequency genome editing using ssDNA oligonucleotides with zinc-finger nucleases. Nat Methods. 2011;8:753–5.

[96] 96. Moehle EA, Rock JM, Lee YL, Jouvenot Y, DeKelver RC, Gregory PD, et al. Targeted gene addition into a specified location in the human genome using designed zinc finger nucleases. Proc Natl Acad Sci U S A. 2007;104:3055–60.

[97] Kuhn AN, Beibetaert T, Simon P, Vallazza B, Buck J, Davies BP, et al. mRNA as a versatile tool for exogenous protein expression. Curr Gene Ther. 2012;12:347–61.

[98] Sahin U, Kariko K, Tureci O. mRNA-based therapeutics—developing a new class of drugs. Nat Rev Drug Discov. 2014;13:759–80.

[99] Skipper KA, Mikkelsen JG. Delivering the goods for genome engineering and editing. Hum Gene Ther. 2015;26:486–97.

[100] Shi NQ, Qi XR, Xiang B, Zhang Y. A survey on "Trojan Horse" peptides: opportunities, issues and controlled entry to "Troy". J Control Release. 2014;194:53–70.

[101] Wadia JS, RVStan, Dowdy SF. Transducible TAT-HA fusogenic peptide enhances escape of TAT-fusion proteins after lipid raft macropinocytosis. Nat Med. 2004;10:310–5.

[102] Gaj T, Guo J, Kato Y, Sirk SJ, Barbas CF, 3rd. Targeted gene knockout by direct delivery of zinc-finger nuclease proteins. Nat Methods. 2012;9:805–7.

[103] Chen Z, Jaafar L, Agyekum DG, Xiao H, Wade MF, Kumaran RI, et al. Receptor-mediated delivery of engineered nucleases for genome modification. Nucleic Acids Res. 2013;41:e182.

[104] Kim S, Kim D, Cho SW, Kim J, Kim JS. Highly efficient RNA-guided genome editing in human cells via delivery of purified Cas9 ribonucleoproteins. Genome Res. 2014;24:1012–9.

[105] Liang X, Potter J, Kumar S, Zou Y, Quintanilla R, Sridharan M, et al. Rapid and highly efficient mammalian cell engineering via Cas9 protein transfection. J Biotechnol. 2015;208:44–53.

[106] Zuris JA, Thompson DB, Shu Y, Guilinger JP, Bessen JL, Hu JH, et al. Cationic lipid-mediated delivery of proteins enables efficient protein-based genome editing *in vitro* and *in vivo*. Nat Biotechnol. 2015;33:73–80.

[107] Papapetrou EP, Zoumbos NC, Athanassiadou A. Genetic modification of hematopoietic stem cells with nonviral systems: past progress and future prospects. Gene Ther. 2005;12 Suppl 1:S118–30.

[108] Rio P, Banos R, Lombardo A, Quintana-Bustamante O, Alvarez L, Garate Z, et al. Targeted gene therapy and cell reprogramming in Fanconi anemia. EMBO Mol Med. 2014;6:835–48.

[109] Perez EE, Wang J, Miller JC, Jouvenot Y, Kim KA, Liu O, et al. Establishment of HIV-1 resistance in $CD4^+$ T cells by genome editing using zinc-finger nucleases. Nat Biotechnol. 2008;26:808–16.

[110] Chira S, Jackson CS, Oprea I, Ozturk F, Pepper MS, Diaconu I, et al. Progresses towards safe and efficient gene therapy vectors. Oncotarget. 2015;6:30675–703.

[111] Chen X, Goncalves MA. Engineered viruses as genome editing devices. Mol Ther. 2015.

[112] Kim H, Kim JS. A guide to genome engineering with programmable nucleases. Nat Rev. 2014;15:321–34.

[113] Liu H, Agarwal S, Kmiec E, Davis BR. Targeted β-globin gene conversion in human hematopoietic CD34(+) and Lin(-)CD38(-)cells. Gene Ther. 2002;9:118–26.

[114] Lombardo A, Genovese P, Beausejour CM, Colleoni S, Lee YL, Kim KA, et al. Gene editing in human stem cells using zinc finger nucleases and integrase-defective lentiviral vector delivery. Nat Biotechnol. 2007;25:1298–306.

[115] Genovese P, Schiroli G, Escobar G, Di Tomaso T, Firrito C, Calabria A, et al. Targeted genome editing in human repopulating haematopoietic stem cells. Nature. 2014;510:235–40.

[116] McNeer NA, Schleifman EB, Cuthbert A, Brehm M, Jackson A, Cheng C, et al. Systemic delivery of triplex-forming PNA and donor DNA by nanoparticles mediates site-specific genome editing of human hematopoietic cells *in vivo*. Gene Ther. 2013;20:658–69.

[117] Hoban MD, Cost GJ, Mendel MC, Romero Z, Kaufman ML, Joglekar AV, et al. Correction of the sickle cell disease mutation in human hematopoietic stem/progenitor cells. Blood. 2015;125:2597–604.

[118] Santilli G, Thrasher AJ. Patching up hematopoietic stem cells. Nat Biotechnol. 2015;33:1236–8.

[119] Wang J, Exline CM, DeClercq JJ, Llewellyn GN, Hayward SB, Li PW, et al. Homology-driven genome editing in hematopoietic stem and progenitor cells using ZFN mRNA and AAV6 donors. Nat Biotechnol. 2015.

[120] Vierstra J, Reik A, Chang KH, Stehling-Sun S, Zhou Y, Hinkley SJ, et al. Functional footprinting of regulatory DNA. Nat Methods. 2015;12:927–30.

[171] Aiuti A, Biasco L, Scaramuzza S, Ferrua F, Cicalese MP, Baricordi C, et al. Lentiviral hematopoietic stem cell gene therapy in patients with Wiskott-Aldrich syndrome. Science. 2013;341:1233151.

[172] Castiello MC, Scaramuzza S, Pala F, Ferrua F, Uva P, Brigida I, et al. B-cell reconstitution after lentiviral vector-mediated gene therapy in patients with Wiskott-Aldrich syndrome. J Allergy Clin Immunol. 2015;136:692–702 e2.

[173] Cavazzana-Calvo M, Payen E, Negre O, Wang G, Hehir K, Fusil F, et al. Transfusion independence and HMGA2 activation after gene therapy of human β-thalassaemia. Nature. 2010;467:318–22.

[174] Cartier N, Hacein-Bey-Abina S, Bartholomae CC, Veres G, Schmidt M, Kutschera I, et al. Hematopoietic stem cell gene therapy with a lentiviral vector in X-linked adrenoleukodystrophy. Science. 2009;326:818–23.

[175] McNeer NA, Chin JY, Schleifman EB, Fields RJ, Glazer PM, Saltzman WM. Nanoparticles deliver triplex-forming PNAs for site-specific genomic recombination in $CD34^+$ human hematopoietic progenitors. Mol Ther. 2011;19:172–80.

[176] Li L, Krymskaya L, Wang J, Henley J, Rao A, Cao LF, et al. Genomic editing of the HIV-1 coreceptor CCR5 in adult hematopoietic stem and progenitor cells using zinc finger nucleases. Mol Ther. 2013;21:1259–69.

[177] Mandal PK, Ferreira LM, Collins R, Meissner TB, Boutwell CL, Friesen M, et al. Efficient ablation of genes in human hematopoietic stem and effector cells using CRISPR/Cas9. Cell Stem Cell. 2014;15:643–52.

[178] Sather BD, Romano Ibarra GS, Sommer K, Curinga G, Hale M, Khan IF, et al. Efficient modification of CCR5 in primary human hematopoietic cells using a megaTAL nuclease and AAV donor template. Sci Transl Med. 2015;7:307ra156.

[179] Liu J, Gaj T, Wallen MC, Barbas CF, 3rd. Improved cell-penetrating zinc-finger nuclease proteins for precision genome engineering. Mol Ther Nucleic Acids. 2015;4:e232.

[180] Torres R, Martin MC, Garcia A, Cigudosa JC, Ramirez JC, Rodriguez-Perales S. Engineering human tumour-associated chromosomal translocations with the RNA-guided CRISPR-Cas9 system. Nat Commun. 2014;5:3964.

Chapter 6

Insulin Gene Therapy for Type 1 Diabetes Mellitus: Unique Challenges Require Innovative Solutions

Andrew M Handorf, Hans W Sollinger and
Tausif Alam

Additional information is available at the end of the chapter

http://dx.doi.org/10.5772/62657

Abstract

Type 1 diabetes mellitus (T1DM) is a disease characterized by chronically elevated blood glucose levels that results from the autoimmune destruction of the insulin-producing β cells of the pancreas. While treatment options exist, they all possess serious limitations. Insulin gene therapy provides a promising alternative aimed at replacing insulin production in native non-β cells. For insulin gene therapy applications to be successful in treating T1DM, a glucose-sensitive organ must be targeted for insulin expression, insulin production must be responsive to ever-changing blood glucose levels, and insulin expression must persist long term. In addition, the amount of insulin production is critical, as too little insulin would lead to poor glucose regulation and too much insulin would induce hypoglycemia, a potentially life-threatening state. Together, insulin gene therapy provides challenges that are absent with other gene therapy applications. In this chapter, we examine the challenges of insulin gene therapy and discuss how the two key components of insulin gene therapy—the insulin expression cassette and the delivery vehicle—can be tailored for the successful treatment of T1DM.

Keywords: gene therapy, type 1 diabetes mellitus, lentivirus, insulin, β cells

1. Introduction

Type 1 diabetes mellitus (T1DM) is an autoimmune disorder whereby the β cells of the pancreas are destroyed. Under physiological conditions, β cells synthesize and secrete insulin in response to changes in blood glucose levels. Insulin, in turn, acts on other cells to promote glucose uptake from the blood and thus lower blood glucose levels to normal. Without sufficient numbers of functional β cells, insulin production becomes inadequate and unable to promptly restore normal blood glucose levels. Over time, chronically elevated blood glucose levels, termed hyperglycemia, cause numerous secondary complications, ultimately leading to widespread tissue and organ damage, as well as increased mortality. According to the National Diabetes Statistics Report, 29.1 million people, or 9.3% of the US population, have diabetes [1], with T1DM accounting for roughly 5% of all diagnosed cases in the adult population. The total economic cost of diabetes is estimated to be $245 billion per year, with T1DM costing an estimated $8–14 billion per year. Thus, diabetes poses a significant burden to our society. Unfortunately, no cure exists for T1DM.

1.1. Current treatment options

While there is currently no cure, several therapies exist to better control blood glucose levels. The most common form of therapy involves use of synthetic insulin, which typically requires multiple insulin injections per day. Unfortunately, this option is cumbersome and unable to restore perfect glucose control. As a result, exogenous insulin therapy delays the onset and reduces the severity of secondary complications but is unable to prevent them [2]. In addition, exogenous insulin therapy can cause hypoglycemia, a potentially life-threatening condition characterized by dangerously low blood glucose levels. The insufficiencies of exogenous insulin therapy arise from the fact that most insulin regimens are unable to mimic normal β cell secretion in response to continually fluctuating blood glucose levels. In an effort to improve upon exogenous insulin therapy and better imitate normal physiology, specialized insulin pumps, dubbed artificial pancreata, have been developed to deliver insulin when needed by continuously monitoring blood glucose levels [3, 4]. Artificial pancreata are able to improve glycemic control, but the implanted glucose sensors inevitably accumulate serum proteins that can compromise the accuracy of glucose measurements and consequently affect the precision of insulin delivered, limiting its long-term effectiveness. Ultimately, exogenous insulin therapies provide a suboptimal therapy for T1DM. Conversely, tighter glucose control can be attained through whole-pancreas transplantation [5], but this therapy is severely hampered by a shortage of donor organs and further complicated by the need for lifelong immunosuppression. Transplantation of pancreatic islets was anticipated to minimize the impact of donor shortage, as islets from one donor could be expanded *ex vivo* to a quantity sufficient for multiple recipients [6], but equivalent successes like those observed with whole-pancreas transplantation have yet to be obtained [7]. Hence, there is clearly a need for effective and broadly applicable treatment options for T1DM.

Gene therapy-based treatment options have emerged as a promising alternative. Gene therapy refers to any technique aimed at using genes to treat or prevent diseases, whether it be through delivery of a functional gene to replace a defective one or through knockdown of dysfunctional genes using gene silencing technologies. For T1DM, three primary gene therapy approaches have been explored to prevent or treat the disease. First, researchers have attempted to prevent the autoimmune destruction of β cells by modifying the immune system. Second, researchers have attempted to generate surrogate β cells from native non-β cells to replace the function of

http://dx.doi.org/10.5772/62657

those cells lost from autoimmune destruction. Third, researchers have taken a more straightforward approach and simply attempted to replace the primary function of β cells—insulin production—to treat T1DM. Not surprisingly, each of these approaches has their own advantages and disadvantages, which we will outline in the following sections.

1.2. Prevention of autoimmune β cell destruction

The most logical therapy for T1DM would be to take preemptive measures and prevent the development of autoimmunity that causes β cell destruction in susceptible individuals. Not surprisingly, many researchers have pursued this strategy as a potential alternative to current treatment options, and to do so, various genetic modifications to both β cells and immune cells have been tested in experimental models of T1DM. First, researchers have sought to induce immune tolerance so that β cell antigens are no longer recognized as foreign. French and colleagues sought to induce immune tolerance by driving intrathymic expression of proinsulin under the control of the major histocompatibility complex (MHC) class II promoter and found that it prevented the onset of T1DM. Similarly, Tian and colleagues transduced autologous bone marrow hematopoietic stem cells of non-obese diabetic (NOD) mice with diabetes-resistant MHC class II I-Aβ chain molecules to examine whether this could prevent the development of diabetes. Indeed, they found that expression of this diabetes-associated allele prevented the development of autoreactive T cells by intrathymic deletion and protected the mice from developing insulitis and diabetes [8]. Second, groups have aimed to modulate the immune system through cytokine-based approaches. For example, gene transfer of transforming growth factor-β and calcitonin gene-related peptide have been shown to prevent the onset of diabetes in NOD mice [9, 10]. Lastly, groups have attempted to modify residual β cells so that they can resist autoimmune destruction, an event that generally occurs through apoptosis. Liu *et al.* overexpressed the antiapoptotic gene, bcl2, in β cells to increase the survival of β cells without affecting their function [11].

Although these preventative options have shown promise, they are hampered by several limitations: (1) These strategies rely on the early detection of diabetes. This is difficult because individuals often do not become symptomatic until they have already lost greater than 80% of their β cells. Thus, efforts to protect the remaining β cells would still leave the patient hyperglycemic. (2) T1DM is a multifactorial disease, making it nearly impossible to accurately predict who from the general population will succumb to the disease [12]. Thus, early intervention can be risky and perhaps even accelerate the progression of the disease. (3) The immune system is highly evolved and its complexities are not well understood. Further, innumerable functional redundancies exist that allow the immune system to compensate for the loss of any single factor or pathway. At our current level of understanding, it seems unlikely that selectively targeting a specific component of the immune system could prevent the autoimmune destruction of remaining β cells. As a result, other gene therapy strategies have been explored.

1.3. Reprogramming non-β cells into β cells

The goal of reprogramming non-β cells into surrogate β cells is to create replacement cells that are as similar to native β cells as possible, including the ability to not only synthesize insulin but also store it within secretory granules and secrete it instantaneously upon elevations in blood glucose levels. The most common way that researchers have done so is by overexpressing β cell-specific transcription factors in non-β cells. Transcription factors are DNA-binding proteins that modulate the rate of transcription of specific genes in a cell type-specific manner. During development, transcription factors play a critical role in executing differentiation programs by driving the expression of cell type-specific genes, and during adulthood, transcription factors are important for maintaining the differentiated status of somatic cells. For β cells, the transcription factors PDX1, NeuroD1, Neurog3, Pax4, Pax6, Nkx6.1, Nkx2.2, and MafA are among the many transcription factors ultimately responsible for directing and/or maintaining the β cell fate.

With the aforementioned knowledge in mind, researchers have attempted to overexpress these transcription factors in a variety of cell types with the hopes of reprogramming them into β cells, including pancreatic exocrine cells [13, 14], keratinocytes [15], hepatic oval stem cells [16], adipose-derived stem cells [17], and hepatocytes. Of these, hepatocytes have been most commonly targeted due to the fact that they are closely related developmentally to β cells and easy to target. PDX1, which regulates early pancreas morphogenesis during development and controls glucose-dependent insulin expression in β cells, has been shown to be indispensable for the conversion of non-β cells into β cells. For example, Ferber *et al.* expressed PDX1 in the livers of diabetic mice and observed insulin expression and secretion, as well as restoration of normoglycemia [18]. Expression of NeuroD1 has also been shown to be a potent inducer of insulin expression in both primary duct cells and hepatocytes [19, 20]. However, ectopic expression of Neurog3 and Nkx6.1, which are also associated with β cell development, was unable to generate surrogate β cells. For example, despite the ability of Neurog3 to activate the persistent expression of NeuroD1, the use of Neurog3 in β cell engineering is not sufficient to generate surrogate β cells [21–25]. However, co-delivery of Neurog3 with PDX1 and MafA was successful in converting pancreatic exocrine cells into surrogate β cells [25]. Similarly, Gefen-Halevi *et al.* overexpressed Nkx6.1 in liver cells but found that it alone was unable to induce insulin expression. However, when co-expressed with PDX1, it promoted reprogramming of liver cells to β cells capable of glucose-stimulated insulin secretion [26]. Together, these studies validate the utility of transcription factor-mediated production of surrogate β cells in animal models and underscore its potential for treatment of T1DM.

Despite these positive findings, the long-term success of reprogramming strategies will rely on the ability of newly formed β cells to evade preexisting autoimmunity. This is particularly relevant given the abundance and diversity of previously identified autoantigens involved in the autoimmune destruction of native β cells [99]. It is expected that the more similar a surrogate β cell is to a native β cell, the more likely it will be targeted by the host's immune system and destroyed. For instance, while Ferber *et al.* found that hepatic expression of PDX1 was able to correct diabetic hyperglycemia, they also found that the mice developed hepatitis and were prone to autoimmunity against the newly formed β cells [18]. While other studies in

NOD mice, a mouse model of autoimmune diabetes provides hope that autoimmunity could be avoided through reprogramming strategies [27], these studies must be carried out for longer periods of time to assess true efficacy. Ultimately, it is likely that these strategies will either require lifelong immunosuppression or selective immunomodulation to ensure the survival of the surrogate β cells.

1.4. Insulin gene therapy

Given the autoimmune nature of T1DM, it may actually be unfavorable to produce surrogate cells that closely resemble native β cells. Since the primary function of β cells is to synthesize and secrete insulin, many groups have taken a humbler approach and simply aimed to restore insulin production in non-β cells, a field known as insulin gene therapy. Although insulin is also an autoantigen known to result in native β cell destruction [28], the theoretical risk of recurring autoimmunity is greatly reduced. Of course, it should be emphasized that β cells are extremely precise in their ability to secrete insulin upon demand, so simply expressing insulin in a non-β cell would likely not be an effective means of treating T1DM. For instance, an individual would not want to be producing high levels of insulin at all times because this would cause hypoglycemic episodes. Thus, to understand whether insulin gene therapy is indeed a viable option to treat T1DM, a clear understanding of the intricacies and aptitude of β cells must first be attained. In the following sections, we will outline what makes a β cell so adept at controlling glycemia and cover what criteria must be met for insulin gene therapy approaches to be successful. We will then present an overview of the field of insulin gene therapy to treat T1DM, with an emphasis on the unique challenges not found in other gene therapy applications. More specifically, we will go into detail about the choice of cell type to target *in vivo*, as most cell types are limited in their ability to adequately sense changing blood glucose levels. We will then discuss key features used within insulin expression cassettes to meet the specific needs of the field, with a particular emphasis on glucose-inducible response elements (GIREs), before discussing the advantages and disadvantages of various viral vectors as they pertain to the field of insulin gene therapy. Lastly, we will present future directions necessary to make insulin gene therapy a clinical reality.

2. Insulin gene therapy

2.1. Features of the β cell

In order to determine whether insulin gene therapy possesses the necessary elegance to be a viable treatment option, it is important to understand the key features of β cells that make them so adept at controlling glycemia and the minimum requirements that absolutely must be met to replace their function. Native β cells possess several important features that together allow them to precisely control blood glucose levels. Specifically, the β cell has the ability to regulate insulin production at the transcriptional, post-transcriptional, translational, and post-translational levels, as well as the ability to store and secrete insulin in a highly regulated fashion in response to glucose. While no individual feature is sufficient in itself to control glycemia, nor specific to β cells, the combination of features makes it a remarkably competent cell for its task.

First and most importantly, β cells have the ability to sense and quickly respond to small changes in circulating glucose levels over a broad range of physiological concentrations (2–20 mM) through concentration-dependent entry and metabolism of glucose. They do so through the activity of glucose transporter-2 (GLUT2) and glucokinase. GLUT2 is a transmembrane protein that enables glucose transport across cell membranes, whereas glucokinase is an enzyme that phosphorylates glucose to initiate its intracellular metabolism. GLUT2 and glucokinase have been dubbed the "glucose sensors" of β cells because they enable β cells to sense glucose over a very broad range of concentrations. They are able to do so due to their high K_m for glucose (~17 and 8 mM, respectively), which allows their activity to vary substantially based on glucose availability [29].

β cells also have the ability to secrete insulin in a precisely regulated fashion in response to elevations in blood glucose levels. β cells do so through their capacity to (1) synthesize and store large quantities of insulin within secretory vesicles and (2) generate secondary stimulus-secretion coupling signals to activate nearly instantaneous insulin vesicle exocytosis. Remarkably, β cells can secrete substantial quantities of insulin within a minute after exposure to elevated glucose. This is a consequence of their metabolic capabilities; β cells possess a unique combination of metabolic enzymes that ultimately leads to the generation of signals capable of altering insulin secretion in response to glucose metabolism. Namely, β cells, but not other mammalian cell types, have negligible lactate dehydrogenase activity while displaying increased pyruvate carboxylase activity, which directs pyruvate almost entirely toward mitochondria for subsequent metabolism via the tricarboxylic acid cycle and oxidative phosphorylation [30]. In so doing, β cells increase their ratio of ATP/ADP and activate ATP-sensitive K^+ channels, a key stimulus leading to insulin vesicle exocytosis.

Furthermore, β cells have remarkable control of insulin biosynthesis at the transcriptional, post-transcriptional, translational, and post-translational levels, with each level of control being regulated by glucose availability. At the transcriptional level, increased glucose levels lead to upregulated insulin expression through the enhanced activity of the transcription factors, PDX1 and MafA [31, 32]. At the post-transcriptional and translational levels, β cells have glucose-responsive mechanisms to modulate insulin mRNA stability and the rate of translation, respectively. This glucose-dependent regulation is primarily governed by an RNA-binding protein known as polypyrimidine tract-binding protein (PTB). Association of PTB with a pyrimidine-rich stretch in the 3′ untranslated region (UTR) of the preproinsulin mRNA has been shown to be responsible for glucose-dependent changes in its stability and rate of translation [33, 34]. In fact, the half-life of preproinsulin mRNA has been shown to increase nearly 3-fold as a result of PTB association [35, 36]. At the post-translational level, β cells possess specific enzymes that allow them to process proinsulin—a precursor form—into fully active insulin. Proinsulin conversion involves removal of two basic pairs of amino acids—the C-peptide—and is mediated by the β cell endoproteases, PC1/3 and PC2, and the exopeptidase, carboxypeptidase-H [37, 38]. Altogether, β cells have a variety of glucose-dependent mechanisms to control insulin output.

β cells further refine insulin biosynthesis and secretion in response to other circulating metabolites. First, the gut produces the peptide hormones, glucose-dependent insulinotropic polypeptide (GIP) and glucagon-like peptide-1, which bind specific receptors found predominantly on β cells to bolster insulin production [39]. Second, specific amino acids and free fatty acids can serve as insulin secretagogues. While some amino acids and free fatty acids can actively promote insulin secretion, most of them simply have a role in amplifying the stimulatory effects of glucose [40]. Third, metabolic stress can induce neuronal signals that influence insulin output [41]. Together, these inputs regulate insulin production and secretion from β cells to more precisely maintain glucose levels within a normal range.

2.2. Requirements of insulin gene therapy

Combining all the remarkable features of the β cell, the end product is a cell with impressive glucose sensitivity, the ability to control insulin biosynthesis at several levels in a glucose-dependent fashion, and the ability to store, process, and secrete insulin almost instantaneously in response to glucose. The β cell is truly impressive. With this knowledge in mind, it would be easy to argue that insulin gene therapy lacks the sophistication necessary to adequately treat T1DM, especially given its relative simplicity. However, it is important to keep in mind that the most commonly used treatment for T1DM currently involves repeated injections of synthetic insulin. While this treatment option holds very little sophistication or biomimicry, it has still proven effective enough to remain a viable option since it was first employed as a medication in 1922. With that being said, any treatment that could provide better glycemic control while averting the cumbersome nature of synthetic insulin therapy would be a noteworthy improvement. So, beyond the ability to produce insulin alone (a feat that could likely be achieved in any cell type using a strong constitutive promoter), what other features of β cells are absolutely critical for the success of insulin gene therapy applications?

Given that glucose is the primary stimulus leading to the production of insulin and subsequent clearance of itself, the impressive glucose sensitivity conferred by GLUT2 and glucokinase is a necessary feature that must be present in an insulin-producing surrogate β cell. Without these proteins, a surrogate β cell would not be able to precisely sense glucose concentration and control insulin output over a broad range of circulating glucose concentrations. In addition to glucose sensitivity, it is also important that the insulin-producing surrogate cell has the ability to respond to changing blood glucose levels by modulating insulin output. In so doing, insulin output could be adjusted in a physiologically relevant manner to better control glycemia. It is likewise important for an insulin-producing cell to have the ability to process proinsulin into insulin, given that proinsulin has less than 10% of the biological activity of fully processed insulin [42]. Lastly, it would be ideal for an insulin-producing cell to have the ability to package, store, and secrete insulin almost instantaneously in response to elevated blood glucose levels. However, it is debatable whether this last feature is indeed critical to the success of insulin gene therapy. Diabetes is such a devastating disease because of the secondary complications that arise as a consequence of sustained hyperglycemia, not repeated episodes of transient hyperglycemia. Thus, as long as the insulin-producing surrogate cell has the ability to produce sufficient quantities of insulin within a reasonable time frame in a glucose-responsive fashion, the difference between near instantaneous insulin secretion or secretion with a few minutes' delay may be less critical. The quantity of insulin secreted must simply be large enough to correct hyperglycemia, but not too large as to cause hypoglycemia.

Thus, a long-lasting treatment for T1DM using insulin gene therapy could be achieved, but several criteria must be considered. First, the appropriate cells must be targeted for insulin production. At a minimum, these cells would need to express the glucose sensors, GLUT2 and glucokinase. Second, insulin transgene expression must be responsive to fluctuating blood glucose levels, being upregulated during hyperglycemia and downregulated during euglycemia. There are a variety of mechanisms available to endow an insulin-producing surrogate β cell with this ability. Third, there must be some mechanism in place for the target cell to process proinsulin into mature insulin. Lastly, an appropriate gene correction tool must be utilized to safely and effectively drive long-term insulin expression.

While meeting these criteria provides several challenges, treating T1DM through insulin gene therapy presents additional challenges not associated with other gene therapy strategies. First, insulin has a relatively short half-life compared to many other proteins being used for gene therapy applications. For instance, the circulating half-life of coagulation factor IX, which is deficient in hemophilia B patients, is estimated to be around 18–25 hours [43], whereas the circulating half-life of insulin has been estimated to be around 4–6 minutes. To compensate for the short circulating half-life of insulin, it is necessary to produce large amounts of insulin either by developing a highly effective gene expression system or by transducing a larger number of cells than other gene therapy applications. This makes the choice of gene delivery system a critical one, as the delivery vehicle must be produced in great abundance and transduce cells efficiently *in vivo*. Second, basal insulin production must be kept low during fasting periods and upregulated only when blood glucose levels become elevated. Other gene therapy applications need only to deliver the therapeutic protein of interest constitutively at low levels to correct the clinical manifestation, owing to greater protein stability and the particular function of the protein. If insulin was expressed constitutively at low levels to satisfy basal metabolic activity, there would be long periods of postprandial hyperglycemia, and if the level of constitutive insulin production was increased to effectively control postprandial glucose levels, there would be a very high possibility of hypoglycemia during fasting periods. Thus, insulin expression cannot be driven by a strong constitutive promoter; it instead must be responsive to fluctuating blood glucose levels. This makes the design of the insulin expression cassette extremely important. It also creates a much narrower therapeutic window for dosing than other gene therapy applications. In the following sections, we will discuss these factors in depth.

2.3. Target cells for insulin gene therapy

At a minimum, the target cells chosen for insulin gene therapy would need to express GLUT2 and glucokinase and have innate mechanisms for glucose-responsiveness, thus giving them the ability to sense and respond to continually fluctuating blood glucose levels. Without them, insulin secretion from surrogate β cells would be far less precise. Besides β cells, the only cells that express both GLUT2 and glucokinase are hepatocytes and cells of the hypothalamus and small intestine. Thus, these cells serve as a nice starting point for targeting of insulin transgene expression. It should be noted, however, that a variety of other cells have been targeted for insulin expression, including skeletal myocytes, fibroblasts, and mesenchymal stem cells. An interesting example is the targeting of skeletal myocytes. Skeletal myocytes do not express GLUT2 or glucokinase. Instead, they express GLUT4 and hexokinase, which each have a higher affinity (i.e. lower K_m) for glucose. To endow skeletal myocytes with enhanced glucose sensitivity, Callejas *et al.* co-expressed insulin with glucokinase. Whereas insulin alone was insufficient to adequately treat T1DM in dogs using skeletal myocytes as surrogate β cells, co-expression of glucokinase with insulin was able to normalize fasting hyperglycemia and accelerate glucose disposal after oral challenge [44]. These findings emphasize the importance of glucose sensitivity in treatment of T1DM.

Intestinal K cells provide a particularly promising cell type for insulin gene therapy applications because they not only express GLUT2 and glucokinase, but they also possess the proinsulin processing enzymes and have the machinery for regulated insulin secretion. Cheung and colleagues exploited these unique advantages to generate insulin-producing surrogate β cells from K cells. To do so, they generated transgenic mice expressing human insulin under control of the GIP promoter, a K cell-specific promoter believed to be regulated by glucose [45]. They found that the GIP promoter was able to target insulin expression to K cells specifically, and the transgenic expression of insulin was effective at promoting normal fasting glucose levels and efficient glucose clearance in response to an oral glucose challenge for up to three months after mice were rendered diabetic with streptozotocin (STZ). However, while their findings hold promise, the translation of this strategy to the clinic will rely on their ability to address the following concerns: (1) The gut is one of the primary hubs for the immune system, and given that insulin itself is an autoantigen responsible for native β cell destruction [46], intestinal K cells may be particularly susceptible to recurring autoimmune attack; long-term protection from autoimmunity must be demonstrated. (2) More importantly, the GIP promoter is not only regulated by glucose intake but also by other sources—most notably fats. Thus, patients receiving this treatment would, at the very least, need to maintain a very strict diet to avoid potentially fatal consequences, like hypoglycemia. Further, another study found that glucose alone does not even regulate insulin secretion when controlled by the GIP promoter [47]. Regulation of the GIP promoter must be more thoroughly examined before moving toward human clinical trials.

The most commonly chosen target cells of insulin gene therapy applications, and the one we have chosen, are hepatocytes. Although hepatocytes do not have the machinery to store insulin within secretory vesicles and secrete it in a regulated fashion, they express GLUT2 and glucokinase and possess a robust capacity to synthesize and secrete proteins constitutively. In addition, hepatocytes are attractive targets for insulin expression because they (1) are closely related to β cells developmentally, (2) play a very important role in glucose homeostasis, and (3) are relatively easy to target. As a result, it has been the most commonly targeted organ for *in vivo* production of insulin-producing surrogate β cells and will be the focus for the remainder of this chapter.

2.4. Expression cassette design

After choosing an appropriate cell type for insulin production, there are several considerations that must be taken into account when designing the insulin expression cassette. Perhaps the first decision that must be made is which promoter to use to drive insulin expression. One of the most commonly used promoters within the field of gene therapy is the cytomegalovirus (CMV) promoter. The CMV promoter is a mammalian promoter from the human cytomegalovirus that drives strong, constitutive transgene expression. While the CMV promoter has been used quite frequently to drive insulin expression for treatment of T1DM, there is one fundamental reason why these studies could never be translated to the clinic: Insulin expression must be responsive to glucose, being upregulated when blood glucose levels rise and downregulated to low levels during fasting periods. The CMV promoter would drive consistent, high level expression of insulin even during fasting periods, which would ultimately cause blood glucose levels to fall dangerously low. Furthermore, even if regulatory elements were added to the expression cassette to endow glucose-responsiveness to insulin expression, the CMV promoter is so strong that it would override these elements. Thus, a weaker promoter must be used to maintain low levels of insulin production during fasting periods if insulin gene therapy is to be successful in treating T1DM.

Weaker tissue-specific promoters have been employed for hepatic insulin gene therapy applications to not only reduce the potential for hypoglycemia but also to improve targeting to the tissue of choice. For instance, several groups have used liver-specific promoters that activate insulin transgene expression in hepatocytes but remain inactive in other cell types. Interestingly, some liver-specific promoters are inherently glucose-responsive, making them a great choice for insulin gene therapy applications. For instance, Chen *et al.* used the glucose-6-phosphatase (G6Pase) promoter to drive insulin expression and found that elevated glucose concentrations enhanced promoter activity [48]. They also found that insulin strongly inhibited G6Pase promoter activity under low glucose conditions, creating a system with feedback inhibition [49]. The group then delivered the insulin gene to the liver of STZ-induced diabetic rats under the control of the G6Pase promoter and found that *ad libitum* hyperglycemia was significantly reduced, glucose utilization was accelerated after glucose challenge, and fasting glucose levels were within a normal range without hypoglycemia. Similarly, Burkhardt *et al.* used the liver-specific GLUT2 promoter to drive insulin gene expression in a glucose-inducible but insulin-repressive fashion and found an improvement in diabetic hyperglycemia [50]. However, it is worth noting that the activity of the wild-type insulin promoter used by native β cells is actually enhanced by insulin, creating a feed-forward system to amplify insulin expression.

To generate a more physiologically mimetic system driving insulin expression, Hsu and coworkers used the rat insulin-1 promoter, creating a system that is activated by both glucose and insulin, similar to native β cells [51]. In so doing, they were successful at driving insulin secretion from Huh7 hepatoma cells *in vitro* in response to glucose. They were also able to augment insulin expression *in vivo* in response to glucose and theophylline—a pharmacological activator of cAMP—and ameliorate hyperglycemia in STZ-induced diabetic mice. However, they did not test whether insulin activated transgene expression using this promoter,

http://dx.doi.org/10.5772/62657

and it is unclear how this promoter was active in hepatocytes, which do not ordinarily express the β cell-specific transcription factors necessary to upregulate insulin expression. Regardless, hepatocytes do not possess enough of the β cell-specific regulatory mechanisms to safely express insulin in a feed-forward manner with respect to insulin. Thus, for the sake of hepatic insulin gene therapy, it would be simpler to create a system that was unresponsive to insulin altogether.

Even if a liver-specific promoter does not possess glucose responsiveness, it can be endowed with sensitivity to glucose through incorporation of GIREs. GIREs are glucose-responsive DNA sequences found in the promoter region of several genes encoding lipogenic enzymes, like L-pyruvate kinase (L-PK), S_{14}, fatty acid synthase, and acetyl-CoA carboxylase [52]. GIREs are composed of two 6-bp motifs known as E boxes, with a consensus sequence of CACGTGnnnnnCACGTG (**Figure 1**). E boxes are generally recognized by transcription factors harboring basic helix-loop-helix/leucine zipper DNA-binding domains [53]. A specific transcription factor, dubbed carbohydrate response element-binding protein (ChREBP), has been found in great abundance in the liver, as well as the small intestine and adipose tissue, the most active sites of *de novo* lipogenesis [54].

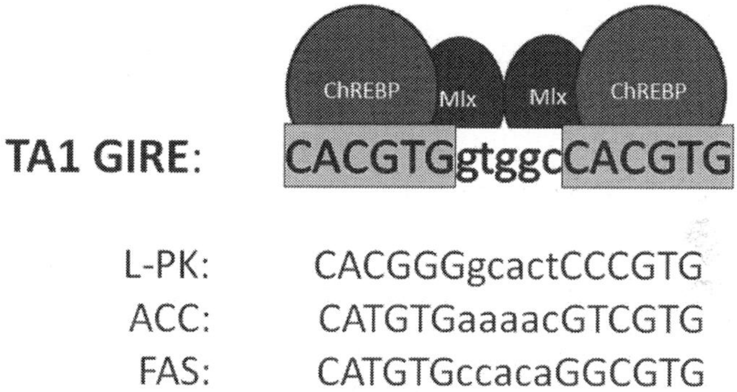

Figure 1. Glucose-inducible response elements (GIREs) and their transcriptional activators.

Incorporation of GIREs enables glucose-responsive control of gene transcription. GIREs are composed of two 6-bp motifs with a consensus sequence of CACGTG separated by 5 bp. A tetramer of ChREBP-Mlx binds each GIRE to amplify gene transcription in response to elevated glucose levels. GIREs have been identified in liver-specific genes like L-pyruvate kinase (L-PK), acetyl-CoA carboxylase (ACC), and fatty acid synthase (FAS).

Thule and colleagues leveraged GIREs to endow glucose-responsiveness to the liver-specific insulin-like growth factor binding protein-1 (IGFBP1) promoter. The IGFBP1 promoter is repressed by insulin, creating a feedback inhibition loop on insulin expression, but it is not inherently influenced by changes in glucose concentrations. To generate glucose-responsive insulin expression, they incorporated GIREs from the L-PK gene directly upstream of the IGFBP1 promoter and found that, depending on the number of GIREs incorporated, a 1.6- to

6.4-fold increase in promoter activity could be produced in response to elevated glucose concentrations in primary hepatocytes *in vitro* [55]. In addition, Thule and Liu used a recombinant adenovirus to deliver their glucose-responsive, insulin-repressive insulin construct into STZ-induced diabetic rats and found that it was able to produce near-normal glycemia and weight gain without inducing lethal hypoglycemia [56].

In our lab, we used the liver-specific albumin promoter—which is neither glucose- nor insulin-responsive—and inserted the GIREs from the S_{14} gene upstream of the albumin promoter to create a system that is unresponsive to insulin but activated by elevated glucose levels (**Figure 2**) [57]. To test the effect that the S_{14} GIREs have on glucose-responsive insulin output from the albumin promoter, we first generated insulin expression cassettes containing one to five GIREs. Interestingly, we found that the degree of glucose-induction on insulin output increased as the number of GIREs was increased up to three, after which there was only a marginal enhancement in insulin output. We observed a 9-fold increase in insulin output from primary hepatocytes between low and high glucose conditions when three GIREs were incorporated upstream of the albumin promoter (**Figure 3**). When we delivered this insulin expression cassette into the livers of STZ-induced diabetic rats through direct injection, we found that fasting blood glucose levels were reduced to normal, blood glucose levels of diabetic rats fed *ad libitum* were significantly reduced, and glucose clearance was significantly accelerated during an intraperitoneal glucose tolerance test. However, these effects only lasted for roughly a month, owing to the use of adenoviral vectors to deliver our insulin expression cassette.

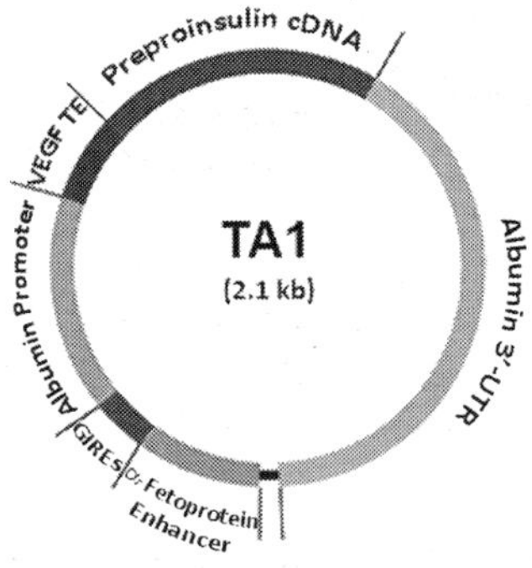

	α-Fetoprotein Enhancer	3x GIREs	Albumin Promoter	VEGF Transl. Enhancer	Insulin (Furin-cleavable)	Albumin 3'-UTR
Liver specificity	+	-	+++	+	-	+
Glucose responsiveness	-	+++	-	-	-	+

Figure 2. Elements of the insulin expression cassette—TA1.

TA1 is a 2.1-kb insulin expression cassette containing elements that drive insulin expression in both a liver-specific and glucose-responsive fashion. The albumin promoter is largely responsible for restricting insulin expression to hepatocytes, while the α-fetoprotein transcriptional enhancer, vascular endothelial growth factor (VEGF) translational enhancer, and albumin 3′-UTR also promote liver specificity. Glucose responsiveness is primarily driven by three copies of GIREs, although the albumin 3′-UTR also promotes glucose-responsive insulin biosynthesis.

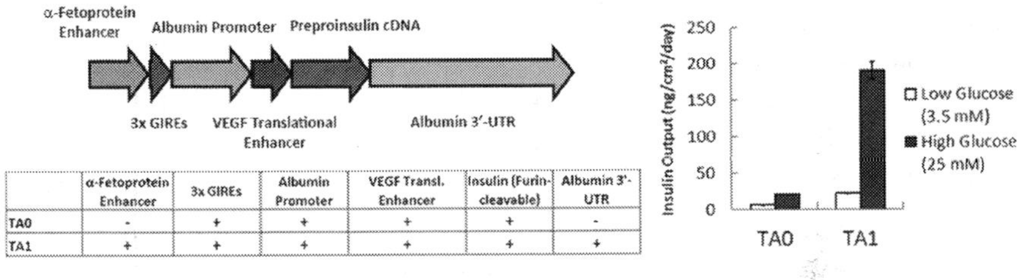

	α-Fetoprotein Enhancer	3x GIREs	Albumin Promoter	VEGF Transl. Enhancer	Insulin (Furin-cleavable)	Albumin 3'-UTR
TA0	-	+	+	+	+	-
TA1	+	+	+	+	+	+

Figure 3. Glucose induction of insulin expression and the effect of hepatocyte-specific enhancer elements on overall insulin output.

The insulin expression cassettes, TA0 and TA1, differ in the presence of α-fetoprotein transcriptional enhancer and the albumin 3′-UTR. Inclusion of these elements greatly increases overall insulin output, while both constructs display similar glucose responsiveness.

To improve our insulin expression cassette, we investigated whether inclusion of various liver-specific enhancer elements could further enhance insulin production. Specifically, we incorporated a transcriptional enhancer from the human α-fetoprotein gene, an intron from the human growth hormone gene previously shown to improve mRNA processing efficiency, a translational enhancer from the vascular endothelial growth factor (VEGF) gene that functions as an internal ribosomal entry site, and the 3′-UTR from the human albumin gene that also contains an intron to improve mRNA processing (**Figure 3**). The ability of each element to enhance glucose-inducible insulin expression was first examined by transducing primary rat hepatocytes *in vitro* using an adenovirus. After testing different constructs containing a variety of different combinations of these elements, we ultimately found that insulin expression cassettes incorporating the α-fetoprotein transcriptional enhancer, VEGF translational enhancer, and albumin 3′-UTR led to increased insulin production *ex vivo*. Specifically, the VEGF translational enhancer led to a 4- to 6-fold increase in insulin output alone at both low and high glucose concentrations. Incorporation of the transcriptional enhancer and 3′-UTR led to another 5- to 8-fold increase in insulin output (**Figure 3**). Together, these modifications to the insulin expression cassette resulted in a 20- to 50-fold increase in insulin output *in vitro* compared to the original constructs, thus allowing us to more efficiently drive insulin expression, a particularly important factor when less efficient delivery vehicles must be used for gene therapy.

We confirmed the utility of this improved insulin construct *in vivo* by delivering our insulin expression cassette in the form of minicircle DNA, which can be readily produced in large quantities. Upon intravenous injection of this minicircle DNA into STZ-induced diabetic rats, we observed a DNA dose-dependent correction in hyperglycemia in both fasted rats and rats fed *ad libitum*. In addition, we observed a full restoration in the rate of weight gain in STZ-induced diabetic rats comparable to that of healthy, non-diabetic rats, and intraperitoneal glucose tolerance tests demonstrated glucose-inducible increases in insulin production capable of correcting hyperglycemia within 45 minutes. A single injection of minicircle DNA led to normalization of serum levels of albumin, triglycerides, cholesterol, aspartate transaminase, alanine aminotransferase, and alkaline phosphatase, thus demonstrating restoration of healthy liver function. Further, there were no signs of hepatic inflammation, underscoring the safety of hepatic insulin gene therapy. Together, we were able to create a treatment for T1DM possessing glucose-responsive insulin production (due to the natural expression of GLUT2 and glucokinase from hepatocytes, and the presence of GIREs in the insulin expression cassette) that is capable of fully correcting diabetic hyperglycemia.

Another novel feature of our insulin expression cassette is the presence of the albumin 3′-UTR. As mentioned previously, the albumin 3′-UTR contains an intron that improves mRNA processing. However, in addition to that, it also contains two pyrimidine-rich stretches known to bind PTB [58]. PTB is a ubiquitously expressed mRNA binding protein that serves as a common mediator of mRNA stability. As mentioned previously, PTB binding sequences are also found in the 3′-UTR of the preproinsulin gene. This is a particularly important feature for hepatic insulin gene therapy applications, as the half-life of preproinsulin mRNA has been reported to be less than 6 hours in hepatocytes. That is much less than the 29–77 hours found in β cells. Perhaps even more importantly, the presence of PTB binding sites would also confer glucose-responsive control of preproinsulin mRNA translation in hepatocytes [34]. Thus, the presence of the albumin 3′-UTR endows hepatocyte-derived surrogate β cells with improved mRNA processing and stability, as well as glucose-responsive control of translation.

One final consideration when designing an expression cassette for insulin gene therapy is the preproinsulin sequence used. A mature insulin molecule is composed of two polypeptide chains—the A and B chains—linked together by two disulfide bonds. However, the insulin gene produces a single preproinsulin polypeptide that contains two basic pairs of amino acids separating the A and B chains, known as the C-peptide, as well as a 24-residue signal peptide. The signal peptide is removed as preproinsulin is translocated into the rough endoplasmic reticulum, forming proinsulin. Proinsulin undergoes further maturation within secretory granules through the action of prohormone convertases PC1/3 and PC2, as well as carboxypeptidase-H. These enzymes are co-packaged with proinsulin in secretory granules and together act to remove C-peptide and produce mature insulin. However, prohormone convertases PC1/3 and PC2 are only found in β cells and other cells with the regulated secretory pathway, like pituitary cells and intestinal K cells. Thus, for insulin gene therapy applications to be successful, it is important to maintain proinsulin processing, even if researchers choose to target cell types that do not have the regulated secretory pathway, like hepatocytes. In these instances, modifications can be made to the preproinsulin sequence to bypass the necessity of PC1/3 and PC2. The most commonly used modification is incorporation of furin cleavage sites [59, 60]. Furin is a ubiquitously expressed endoprotease that can efficiently cleave proteins at paired basic amino acid sites. Through incorporation of furin cleavage sites, any cell of the body can produce fully functional insulin.

Further modifications can be made to the preproinsulin sequence to enhance bioactivity or production for insulin gene therapy applications. First, the preproinsulin sequence can be mutated to alter the stability of the resulting insulin molecules. The most prevalently used mutation is the B10 mutation—a naturally occurring mutation where the histidine residue at position 10 on the B chain is replaced by aspartic acid [59, 61]. This mutation results in enhanced stability and the accumulation of 10- to 100-fold more mature insulin than wild-type insulin. Other mutations have been found to result in highly potent insulin analogues, including HisA8, ArgA8, and GluB10 [61].

Another way the preproinsulin sequence can be modified is through codon optimization. A codon is a series of three nucleotides that encode a specific amino acid. There are 64 different codons but only 20 different amino acids, which means that many amino acids are encoded by multiple codons. It is generally acknowledged that different organisms have codon preferences as a result of the composition of their respective tRNA pool. In other words, specific codons are preferred by specific organisms because they have that specific tRNA in greater abundance. It is thought that cDNA sequences with optimized codons will achieve faster rates of translation and accuracy, thus improving translational efficiency and production of the transgene product. For gene therapy applications, this has been shown to improve the potency of treatment. For instance, Cantore *et al.* observed a 2- to 3-fold increase in potency of their factor IX treatment for hemophilia B in dogs upon codon optimization [62]. Codon-optimized versions of human insulin have also been shown to achieve better glycemic control in diabetic dogs due to increases in insulin production [44].

In summary, there is some flexibility in the design of the expression cassette for insulin gene therapy applications; the relatively small size of the preproinsulin gene is advantageous for the design of an expression cassette and its subsequent delivery to target cells. Regulatory elements capable of improving cell type specificity, overall insulin output, and glucose responsiveness can be employed to yield insulin expression with greater precision. In addition, the preproinsulin sequence itself can be modified to improve production and functionality. Once a sufficient level of control has been attained over insulin expression, it then becomes a matter of delivering the expression cassette to target cells in a safe and efficient manner.

2.5. Delivery vehicles for insulin gene therapy

When choosing a delivery vehicle for insulin gene therapy applications, two important considerations must be taken into account. First, the delivery vehicle must be able to promote long-term insulin expression. This is important because an antibody response from the initial treatment will reduce the efficacy of subsequent treatments. Thus, repeated administration of most delivery vehicles is largely ineffective, especially if it occurs more than two weeks after the initial treatment. Second, it must be possible to affordably and reliably produce the gene delivery vehicle in the large quantities needed for gene therapy. This is particularly important in the field of insulin gene therapy because the insulin molecule has a relatively short circulating half-life, estimated at around 4–6 minutes [63]. As a result, a greater number of cells must be targeted for insulin expression than other gene therapy applications.

Many gene delivery vehicles exist and can be broadly grouped into two categories: viral and non-viral. Non-viral methods have the advantages of being safer and inducing less of an immune response. However, non-viral methods typically only drive transgene expression transiently, as most are incapable of supporting chromosomal integration. In addition, they tend to deliver genes inefficiently *in vivo*. Regardless, we explored the potential of delivering the insulin gene in the form of minicircle DNA to validate the *in vivo* efficacy of our expression cassette. We chose to use minicircle DNA because they can be produced in large quantities and contain no bacterial or viral elements, improving their likelihood of evading the immune system. We found that delivering our insulin expression cassette as a minicircle was able to correct diabetic hyperglycemia in a dose-dependent fashion in STZ-induced diabetic rats (**Figure 4**). While the effects of this treatment only persisted for about a month, the reduced immunogenicity of minicircle DNA allowed for repeated administration, although the second injection was not nearly as effective as the first (**Figure 5**) [64].

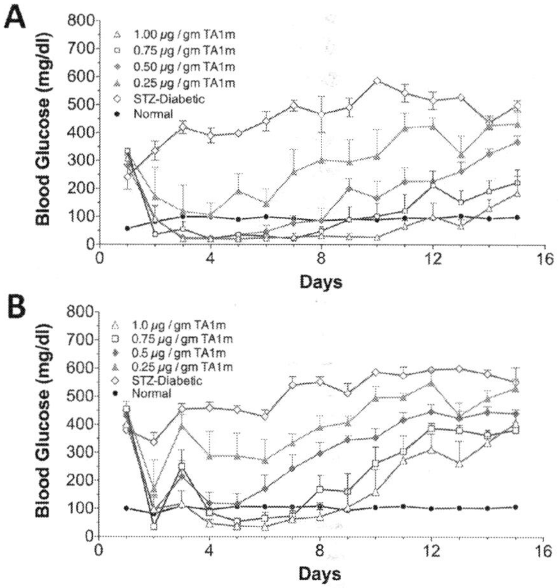

Figure 4. Dose-dependent effect of insulin minicircle DNA treatment on hyperglycemia in rats.

STZ-induced diabetic rats were intravenously injected with the indicated dose of TA1m minicircle DNA and measured for both fasting (A) and *ad libitum* (B) blood glucose levels. There was a dose-dependent correction of diabetic hyperglycemia that persisted for at least 10 days.

http://dx.doi.org/10.5772/62657

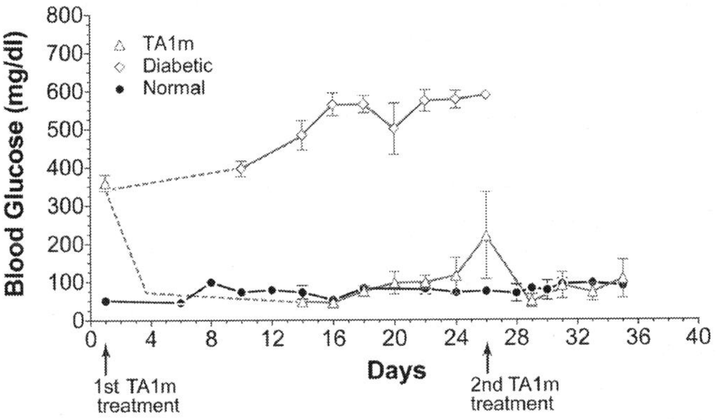

Figure 5. Effect of repeated administration of insulin minicircle DNA on fasting hyperglycemia in diabetic rats.

STZ-induced diabetic rats were intravenously injected with 0.8 μg/gm body weight of TA1m via tail vein, and blood glucose measurements were made after a 4-hour fast. TA1m was able to correct diabetic hyperglycemia for nearly a month before the effects began to diminish. A second TA1m injection (0.8 μg/gm body weight) was made on the 26th day and was able to re-correct fasting blood glucose levels, thus demonstrating that a substantial humoral response had not been mounted against the minicircle DNA.

We have also explored the use of viral delivery vehicles, as viruses are highly evolved and proficient at delivering genetic information into target cells. Of course, viral vectors will inevitably elicit an immune response, with some viruses invoking a greater immune response than others. Several viral vectors have been used for insulin gene therapy, with each possessing their own unique features. Refer to **Table 1** for an overview of the features inherent to various viral vectors.

Viral vector	Packaging capacity	Length of expression	Relative viral titer	Transduction efficiency	Infect both dividing and non-dividing cells	Immunogenicity
Adenovirus	7.5 kb	Transient	+++	+++	Yes	High
Adeno-associated virus	4.5 kb	Transient and Stable	++	++	Yes	Low
Oncoretrovirus	8 kb	Stable	+	+	No	Moderate
Lentivirus	8 kb	Stable	+	++	Yes	Low

Table 1. Features of various viral vectors for gene therapy applications.

Adenovirus, adeno-associated virus, oncoretrovirus, and lentivirus are the most commonly used delivery vehicles for gene therapy applications. Each viral vector possesses their own unique features that affects their suitably for insulin gene therapy applications.

Adenoviruses were among the first viral vectors used in the field of insulin gene therapy due to their ability to be produced in very high titers and transduce non-dividing cells with high efficiency. These features allow researchers to transduce a large number of cells *in vivo* and overcome the lack of insulin protein stability. However, adenoviral vectors are unable to integrate their genetic cargo into the host's genome and thus provide only transient gene expression [65, 66]. As a proof-of-principle, we initially used adenoviral vectors to establish the efficacy of our insulin expression cassette *in vivo*. Indeed, we were successful at correcting fasting blood glucose levels and improving *ad libitum* glucose levels in diabetic rats (**Figure 6**). As expected, however, the observed reduction in blood glucose levels only persisted for about a month. This observation is in agreement with other studies using adenoviral vectors to deliver the insulin gene into diabetic animals, which likewise noted an improvement of hyperglycemia for only a month [67]. Unfortunately, unlike minicircle DNA, a humoral response is elicited by the first treatment, precluding repeated administration of adenovirus-based treatments. Thus, while adenoviral vectors are extremely efficient gene delivery tools, they are not well suited as a long-term therapeutic tool for treatment of T1DM, at least in their initial form.

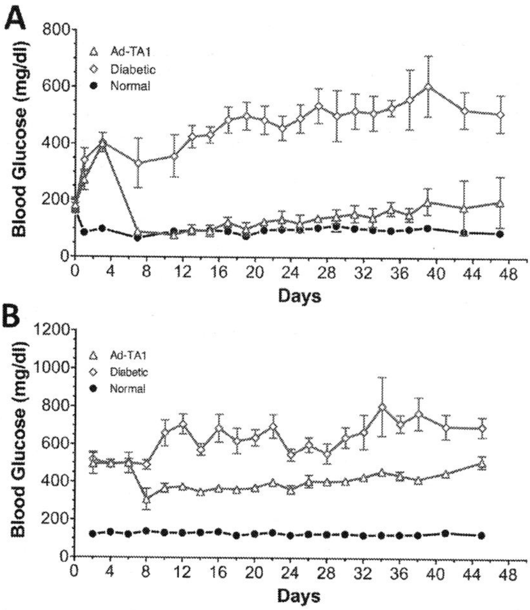

Figure 6. Effect of TA1 on hyperglycemia in diabetic rats when delivered via adenovirus.

STZ-induced diabetic rats were injected with 2×10^{10} adenoviral pfu/rat, and both fasting (A) and *ad libitum* (B) blood glucose measurements were made. Treatment with adenoviral vectors carrying the TA1 insulin expression cassette were able to fully correct diabetic hyperglycemia after an overnight fast and partially correct *ad libitum* blood glucose levels for around one month.

More recently, researchers have generated a newer version of "gutted" adenoviral vector that has been stripped of all viral coding sequences, greatly reducing their immunogenicity [68, 69]. This is an important advancement in the field of gene therapy because adenoviral vectors have proven so immunogenic in past human clinical trials that their administration led to the death of a patient in 1999, temporarily halting progress in the field of gene therapy [70, 71]. Immunogenicity is undoubtedly a very large concern with adenoviral vectors, so any improvement is useful. However, it seems unlikely that immunogenicity will ever be completely eliminated from adenoviral vectors, or any viral vector for that matter. Further advancements have now made it possible to improve upon the innate capabilities of adenoviral vectors by adding the potential to integrate their genetic cargo into a host's genome and drive long-term transgene expression. The advancement was a result of the merging of two technologies, where chromosomal integration is mediated by the Sleeping Beauty (SB) transposon system and efficient gene delivery accomplished by the gutted adenoviral vectors. DNA transposons translocate from one DNA site to another through a simple cutting-and-pasting process, enabling the integration of defined DNA sequences into mammalian genomes [72]. To achieve stable transgene expression using this system, two separate adenoviral vectors must be administered and co-transduce a single cell. The first vector represents the transposon donor vector, which contains a transposon encoding the transgene of interest. The second vector encodes the SB transposase, which mediates relocation. This system has been used successfully to enable persistent phenotypic correction of hemophilia B in dogs [72] and holds great potential for the treatment of other diseases that require persistent gene expression.

To combat issues related to immunogenicity and short-term expression, other groups have employed adeno-associated virus (AAVs). AAVs are able to transduce both dividing and non-dividing cells. In dividing cells, AAVs are able to integrate transgenes into the host's genome at a specific site on chromosome 19 [73]. Within non-dividing cells, the AAV genetic cargo remains largely episomal, as the chromatin is less accessible. AAVs are less immunogenic than adenoviral vectors and are reported to cause relatively long-term transgene expression in non-dividing cells. The primary disadvantage of using AAVs is that their packaging capacity is less than 5 kb, limiting the use of larger expression cassettes with greater complexity for regulated expression. However, the preproinsulin gene is quite small, so even when the gene is accompanied with multiple regulatory elements in a complex expression cassette, the maximum size limitation is unlikely to become an issue for insulin gene therapy applications. Indeed, AAVs have been used to successfully drive insulin expression within non-β cells. Park *et al.* used AAV to deliver insulin under control of the CMV promoter into STZ-induced diabetic rats and found improved glucose tolerance (at 2 g/kg) comparable to that of non-diabetic control rats [74]. Additionally, they observed a less pronounced immune response using AAV when compared to the same treatment using adenoviral vectors.

Retroviral vectors are another widely used gene delivery vehicle, owing to their ability to integrate their genetic cargo into a host's genome and attain sustained gene expression. However, retroviral vectors are greatly limited by their inability to integrate their cargo into the chromosomes of non-dividing cells, a problem that severely hinders their utility in insulin gene therapy applications. In cases where retroviral vectors have been used to deliver insulin, hepatocyte proliferation must first be stimulated [75]. This, of course, greatly limits the translation of retroviral vectors for treatment of T1DM. Retroviral vectors have also been shown to have a preference to integrate their genetic cargo in close proximity to the transcriptional regulatory sequences of proto-oncogenes, as observed in 1999 following the treatment of nine severe-combined immunodeficiency patients [76]. Insertional mutagenesis led to the development of leukemia in four of the nine patients, ultimately halting the field of clinical gene therapy temporarily. All in all, retroviral vectors do not possess favorable features for insulin gene therapy applications.

Lentiviral vectors are a type of retrovirus that provide two key advantages over other retroviruses: (1) they are able to integrate their genetic cargo into the genome of both dividing and non-dividing cells and (2) have less preference to integrate near regulatory sequences of proto-oncogenes, reducing the risk of insertional mutagenesis [77, 78]. Their ability to transduce non-dividing cells is critical for gene therapy strategies, as most cells of the body are either non-dividing or slowly dividing. An additional advantage of lentiviral vectors is that they do not elicit a strong immune response. Unfortunately, lentiviral vectors possess two major pitfalls limiting their widespread translation to human clinical trials: (1) Lentiviral vectors are difficult to produce in high titer [79] and (2) the efficiency of lentiviral transduction *in vivo* is significantly lower than other vectors, especially adenoviral vectors. Given the relative instability of insulin, the need to transduce a larger number of cells than other gene therapy applications, and the fact that—unlike other diseases—a partial correction of hyperglycemia is not sufficient to adequately treat T1DM, these pitfalls pose some limitations to the use of lentiviral vectors for insulin gene therapy applications. Nonetheless, lentiviral vectors offer long-term transgene expression with reduced immunogenicity and an improved biosafety profile and are thus a viable candidate as a therapeutic tool for treatment of T1DM.

To combat issues related to lentivirus infectivity, researchers have modified lentiviral vectors to improve their *in vivo* efficacy. For example, Naldini *et al.* has previously shown that inclusion of viral protein R—a viral protein present in native HIV-1 particles—within synthetic lentiviral particles is critical for hepatocyte transduction [80]. Conversely, Schaffer *et al.* has generated vesicular stromatitis virus envelope proteins that show improved serum resistance [81], which could improve lentiviral efficacy *in vivo*. A combination of modifications will hopefully yield a more potent lentiviral particle *in vivo*. Interestingly, Ren and colleagues used lentiviral vectors to deliver insulin to the livers of STZ-induced diabetic rats [82] and NOD mice [83] and observed long-term correction of diabetic hyperglycemia with no evidence of impaired liver function, intrahepatic inflammation, or recurring autoimmunity against the newly formed insulin-producing cells. While their work gives validity to the use of lentiviral vectors for insulin gene therapy applications, it should be noted that insulin expression was driven by the CMV promoter and displayed no responsiveness to circulating glucose levels.

3. Concluding remarks

Overall, insulin gene therapy provides a promising alternative to current treatments for T1DM. Although this treatment option will inevitably lack the full sophistication of native β cells, it would certainly improve upon current treatment options. Insulin gene therapy opens the possibility of having a one-time treatment option that can provide long-term correction of diabetic hyperglycemia through physiologically relevant mechanisms, like glucose-dependent alterations in insulin transcription and translation. Further, the simplicity of the treatment should yield reproducible results with excellent success rates and additionally help newly-formed insulin-producing surrogate β cells evade recurring autoimmunity. However, several hurdles must still be overcome.

In order for the treatment to be a viable option, long-term insulin expression must be driven to avoid repeated injection. However, the viral vectors currently used to drive long-term transgene expression, like lentivirus, are generally difficult to produce in high titer and limited by their transduction efficiency. As a result, most successful long-term efforts in the field have employed the CMV promoter, which can only drive strong constitutive expression of insulin. In order to take those research efforts to the next level, weaker tissue-specific promoters must be used that drive low basal levels of insulin expression during fasting periods and substantially upregulated insulin expression upon increases in glucose availability. To date, this has yet to be achieved. We are currently exploring several viral vectors for their capacity to deliver our insulin expression cassette—which has elements to drive liver-specific, glucose-responsive insulin expression —at a therapeutic level. In so doing, we hope to produce an affordable, long-term treatment option for patients with T1DM.

Author details

Andrew M Handorf, Hans W Sollinger* and Tausif Alam

*Address all correspondence to: hans@surgery.wisc.edu

University of Wisconsin-Madison, Madison, WI, USA

References

[1] Vehik K, Ajami NJ, Hadley D, Petrosino JF, Burkhardt BR. The changing landscape of type 1 diabetes: recent developments and future frontiers. Current Diabetes Reports. 2013;13(5):642–50. PubMed PMID: 23912764. Pubmed Central PMCID: 3827778.

[2] The Diabetes Control and Complications Trial Research Group. The effect of intensive treatment of diabetes on the development and progression of long-term complications

in insulin-dependent diabetes mellitus. The New England Journal of Medicine. 1993;329(14):977–86. PubMed PMID: 8366922.

[3] Hovorka R. Closed-loop insulin delivery: from bench to clinical practice. Nature Reviews Endocrinology. 2011;7(7):385–95. PubMed PMID: 21343892.

[4] Hovorka R, Kumareswaran K, Harris J, Allen JM, Elleri D, Xing D, et al. Overnight closed loop insulin delivery (artificial pancreas) in adults with type 1 diabetes: crossover randomised controlled studies. BMJ. 2011;342:d1855. PubMed PMID: 21493665. Pubmed Central PMCID: 3077739.

[5] Kelly WD, Lillehei RC, Merkel FK, Idezuki Y, Goetz FC. Allotransplantation of pancreas and duodenum along with kidney in diabetic nephropathy. Surgery. 1967;61(6):827. PubMed PMID: WOS:A19679431400001.

[6] Paty BW, Ryan EA, Shapiro AM, Lakey JR, Robertson RP. Intrahepatic islet transplantation in type 1 diabetic patients does not restore hypoglycemic hormonal counterregulation or symptom recognition after insulin independence. Diabetes. 2002;51(12): 3428–34. PubMed PMID: 12453896.

[7] Shapiro AM, Lakey JR, Ryan EA, Korbutt GS, Toth E, Warnock GL, et al. Islet transplantation in seven patients with type 1 diabetes mellitus using a glucocorticoid-free immunosuppressive regimen. The New England Journal of Medicine. 2000;343(4):230–8. PubMed PMID: 10911004.

[8] Tian CR, Bagley J, Cretin N, Seth N, Wucherpfennig KW, Iacomini J. Prevention of type 1 diabetes by gene therapy. Journal of Clinical Investigation. 2004;114(7):969–78. PubMed PMID: WOS:000224223800018.

[9] Piccirillo CA, Chang Y, Prud'homme GJ. TGF-beta1 somatic gene therapy prevents autoimmune disease in nonobese diabetic mice. Journal of Immunology. 1998;161(8): 3950–6. PubMed PMID: 9780163.

[10] Khachatryan A, Guerder S, Palluault F, Cote G, Solimena M, Valentijn K, et al. Targeted expression of the neuropeptide calcitonin gene-related peptide to beta cells prevents diabetes in NOD mice. Journal of Immunology. 1997;158(3):1409–16. PubMed PMID: 9013986.

[11] Liu Y, Rabinovitch A, Suarez-Pinzon W, Muhkerjee B, Brownlee M, Edelstein D, et al. Expression of the bcl-2 gene from a defective HSV-1 amplicon vector protects pancreatic beta-cells from apoptosis. Human Gene Therapy. 1996;7(14):1719–26. PubMed PMID: 8886842.

[12] Atkinson MA, Maclaren NK. The pathogenesis of insulin-dependent diabetes mellitus. The New England Journal of Medicine. 1994;331(21):1428–36. PubMed PMID: 7969282.

[13] Zhou Q, Brown J, Kanarek A, Rajagopal J, Melton DA. In vivo reprogramming of adult pancreatic exocrine cells to beta-cells. Nature. 2008;455(7213):627–32. PubMed PMID: 18754011.

[14] Li W, Nakanishi M, Zumsteg A, Shear M, Wright C, Melton DA, et al. In vivo reprogramming of pancreatic acinar cells to three islet endocrine subtypes. eLife. 2014;3:e01846. PubMed PMID: 24714494. Pubmed Central PMCID: 3977343.

[15] Mauda-Havakuk M, Litichever N, Chernichovski E, Nakar O, Winkler E, Mazkereth R, et al. Ectopic PDX-1 expression directly reprograms human keratinocytes along pancreatic insulin-producing cells fate. PLoS One. 2011;6(10):e26298. PubMed PMID: 22028850. Pubmed Central PMCID: 3196540.

[16] Li H, Li X, Lam KS, Tam S, Xiao W, Xu R. Adeno-associated virus-mediated pancreatic and duodenal homeobox gene-1 expression enhanced differentiation of hepatic oval stem cells to insulin-producing cells in diabetic rats. Journal of Biomedical Science. 2008;15(4):487–97. PubMed PMID: 18253862.

[17] Lin G, Wang G, Liu G, Yang LJ, Chang LJ, Lue TF, et al. Treatment of type 1 diabetes with adipose tissue-derived stem cells expressing pancreatic duodenal homeobox 1. Stem cells and development. 2009;18(10):1399–406. PubMed PMID: 19245309. Pubmed Central PMCID: 2862049.

[18] Ferber S, Halkin A, Cohen H, Ber I, Einav Y, Goldberg I, et al. Pancreatic and duodenal homeobox gene 1 induces expression of insulin genes in liver and ameliorates streptozotocin-induced hyperglycemia. Nature Medicine. 2000;6(5):568–72. PubMed PMID: 10802714.

[19] Noguchi H, Xu G, Matsumoto S, Kaneto H, Kobayashi N, Bonner-Weir S, et al. Induction of pancreatic stem/progenitor cells into insulin-producing cells by adenoviral-mediated gene transfer technology. Cell Transplant. 2006;15(10):929–38. PubMed PMID: WOS:000243780000012. English.

[20] Yatoh S, Akashi T, Chan PP, Kaneto H, Sharma A, Bonner-Weir S, et al. NeuroD and reaggregation induce beta-cell specific gene expression in cultured hepatocytes. Diabetes/Metabolism Research and Reviews. 2007;23(3):239–49. PubMed PMID: 16921545.

[21] Wang AY, Ehrhardt A, Xu H, Kay MA. Adenovirus transduction is required for the correction of diabetes using Pdx-1 or neurogenin-3 in the liver. Molecular Therapy. 2007;15(2):255–63. PubMed PMID: WOS:000244405200007.

[22] Kaneto H, Nakatani Y, Miyatsuka T, Matsuoka T, Matsuhisa M, Hori M, et al. PDX-1/VP16 fusion protein, together with NeuroD or Ngn3, markedly induces insulin gene transcription and ameliorates glucose tolerance. Diabetes. 2005;54(4):1009–22. PubMed PMID: WOS:000228094300012.

[23] Song YD, Lee EJ, Yashar P, Pfaff LE, Kim SY, Jameson JL. Islet cell differentiation in liver by combinatorial expression of transcription factors Neurogenin-3, BETA2, and RIPE3b1. Biochemical and Biophysical Research Communications. 2007;354(2):334–9. PubMed PMID: WOS:000244050000002.

[24] Heremans Y, Van De Casteele M, Veld PI, Gradwohl G, Serup P, Madsen O, et al. Recapitulation of embryonic neuroendocrine differentiation in adult human pancreatic duct cells expressing neurogenin 3. J Cell Biol. 2002;159(2):303–11. PubMed PMID: WOS:000179051500011.

[25] Zhou Q, Brown J, Kanarek A, Rajagopal J, Melton DA. In vivo reprogramming of adult pancreatic exocrine cells to beta-cells. Nature. 2008;455(7213):627–30. PubMed PMID: WOS:000259639700038.

[26] Gefen-Halevi S, Rachmut IH, Molakandov K, Berneman D, Mor E, Meivar-Levy I, et al. NKX6.1 promotes PDX-1-induced liver to pancreatic beta-cells reprogramming. Cellular Reprogramming. 2010;12(6):655–64. PubMed PMID: 21108535.

[27] Shternhall-Ron K, Quintana FJ, Perl S, Meivar-Levy I, Barshack I, Cohen IR, et al. Ectopic PDX-1 expression in liver ameliorates type 1 diabetes. Journal of Autoimmunity. 2007;28(2–3):134–42. PubMed PMID: 17383157.

[28] Roep BO, Peakman M. Antigen targets of type 1 diabetes autoimmunity. Cold Spring Harbor Perspectives in Medicine. 2012;2(4):a007781. PubMed PMID: 22474615. Pubmed Central PMCID: 3312399.

[29] Newgard CB, McGarry JD. Metabolic coupling factors in pancreatic beta-cell signal transduction. Annual Review of Biochemistry. 1995;64:689–719. PubMed PMID: 7574498.

[30] Ishihara H, Wang H, Drewes LR, Wollheim CB. Overexpression of monocarboxylate transporter and lactate dehydrogenase alters insulin secretory responses to pyruvate and lactate in beta cells. The Journal of Clinical Investigation. 1999;104(11):1621–9. PubMed PMID: 10587526. Pubmed Central PMCID: 409861.

[31] Melloul D, Marshak S, Cerasi E. Regulation of insulin gene transcription. Diabetologia. 2002;45(3):309–26. PubMed PMID: 11914736.

[32] Zhao L, Guo M, Matsuoka TA, Hagman DK, Parazzoli SD, Poitout V, et al. The islet beta cell-enriched MafA activator is a key regulator of insulin gene transcription. The Journal of Biological Chemistry. 2005;280(12):11887–94. PubMed PMID: 15665000.

[33] Tillmar L, Welsh N. Glucose-induced binding of the polypyrimidine tract-binding protein (PTB) to the 3′-untranslated region of the insulin mRNA (ins-PRS) is inhibited by rapamycin. Molecular and Cellular Biochemistry. 2004;260(1–2):85–90. PubMed PMID: WOS:000221121000011.

[34] Wicksteed B, Herbert TP, Alarcon C, Lingohr MK, Moss LG, Rhodes CJ. Cooperativity between the preproinsulin mRNA untranslated regions is necessary for glucose-stimulated translation. Journal of Biological Chemistry. 2001;276(25):22553–8. PubMed PMID: WOS:000169412700078.

[35] Welsh M, Nielsen DA, MacKrell AJ, Steiner DF. Control of insulin gene expression in pancreatic beta-cells and in an insulin-producing cell line, RIN-5F cells. II. Regulation

of insulin mRNA stability. The Journal of Biological Chemistry. 1985;260(25):13590–4. PubMed PMID: 3902821.

[36] Tillmar L, Welsh N. Hypoxia may increase rat insulin mRNA levels by promoting binding of the polypyrimidine tract-binding protein (PTB) to the pyrimidine-rich insulin mRNA 3′-untranslated region. Mol Med. 2002;8(5):263–72. PubMed PMID: WOS:000178018100005.

[37] Hutton JC. Insulin secretory granule biogenesis and the proinsulin-processing endopeptidases. Diabetologia. 1994;37:S48–56. PubMed PMID: WOS:A1994PE89100008. English.

[38] Steiner DF. The proprotein convertases. Current Opinion in Chemical Biology. 1998;2(1):31–9. PubMed PMID: WOS:000072701000005.

[39] McIntosh CHS, Widenmaier S, Kim SJ. Glucose-dependent insulinotropic polypeptide (Gastric Inhibitory Polypeptide; Gip). Vitamins & Hormones. 2009;80:409–71. PubMed PMID: WOS:000264638200015.

[40] Nolan CJ, Prentki M. The islet beta-cell: fuel responsive and vulnerable. Trends in Endocrinology & Metabolism. 2008;19(8):285–91. PubMed PMID: WOS: 000260444900003.

[41] Pongratz RL, Kibbey RG, Shulman GI, Cline GW. Cytosolic and mitochondrial malic enzyme isoforms differentially control insulin secretion. The Journal of Biological Chemistry. 2007;282(1):200–7. PubMed PMID: 17102138.

[42] Revers RR, Henry R, Schmeiser L, Kolterman O, Cohen R, Bergenstal R, et al. The effects of biosynthetic human proinsulin on carbohydrate-metabolism. Diabetes. 1984;33(8): 762–70. PubMed PMID: WOS:A1984TB55400010.

[43] Peters RT, Low SC, Kamphaus GD, Dumont JA, Amari JV, Lu Q, et al. Prolonged activity of factor IX as a monomeric Fc fusion protein. Blood. 2010;115(10):2057–64. PubMed PMID: 20056791.

[44] Callejas D, Mann CJ, Ayuso E, Lage R, Grifoll I, Roca C, et al. Treatment of diabetes and long-term survival after insulin and glucokinase gene therapy. Diabetes. 2013;62(5): 1718–29. PubMed PMID: 23378612. Pubmed Central PMCID: 3636629.

[45] Cheung AT, Dayanandan B, Lewis JT, Korbutt GS, Rajotte RV, Bryer-Ash M, et al. Glucose-dependent insulin release from genetically engineered K cells. Science. 2000;290(5498):1959–62. PubMed PMID: WOS:000165810000044.

[46] Nakayama M. Insulin as a key autoantigen in the development of type 1 diabetes. Diabetes/Metabolism Research. 2011;27(8):773–7. PubMed PMID: WOS: 000300108000009.

[47] Ramshur EB, Rull TR, Wice BM. Novel insulin/GIP co-producing cell lines provide unexpected insights into Gut K-cell function in vivo. Journal of Cellular Physiology. 2002;192(3):339–50. PubMed PMID: 12124779.

[48] Chen RH, Meseck ML, Woo SLC. Auto-regulated hepatic insulin gene expression in type 1 diabetic rats. Molecular Therapy. 2001;3(4):584–90. PubMed PMID: WOS: 000168367500023.

[49] Chen R, Meseck M, McEvoy RC, Woo SL. Glucose-stimulated and self-limiting insulin production by glucose 6-phosphatase promoter driven insulin expression in hepatoma cells. Gene Therapy. 2000;7(21):1802–9. PubMed PMID: 11110411.

[50] Burkhardt BR, Parker MJ, Zhang YC, Song SH, Wasserfall CH, Atkinson MA. Glucose transporter-2 (GLUT2) promoter mediated transgenic insulin production reduces hyperglycemia in diabetic mice. FEBS Letters. 2005;579(25):5759–64. PubMed PMID: WOS:000232824200053.

[51] Hsu PYJ, Kotin RM, Yang YW. Glucose- and metabolically regulated hepatic insulin gene therapy for diabetes. Pharmaceutical Research. 2008;25(6):1460–8. PubMed PMID: WOS:000255753900022.

[52] Towle HC. Glucose as a regulator of eukaryotic gene transcription. Trends in Endocrinology & Metabolism. 2005;16(10):489–94. PubMed PMID: WOS:000234024200006.

[53] Atchley WR, Fitch WM. A natural classification of the basic helix-loop-helix class of transcription factors. Proceedings of the National Academy of Sciences of the United States of America. 1997;94(10):5172–6. PubMed PMID: 9144210. Pubmed Central PMCID: 24651.

[54] Iizuka K, Bruick RK, Liang G, Horton JD, Uyeda K. Deficiency of carbohydrate response element-binding protein (ChREBP) reduces lipogenesis as well as glycolysis. Proceedings of the National Academy of Sciences of the United States of America. 2004;101(19): 7281–6. PubMed PMID: 15118080. Pubmed Central PMCID: 409910.

[55] Thule PM, Liu J, Phillips LS. Glucose regulated production of human insulin in rat hepatocytes. Gene Therapy. 2000;7(3):205–14. PubMed PMID: 10694797.

[56] Thule PM, Liu JM. Regulated hepatic insulin gene therapy of STZ-diabetic rats. Gene Therapy. 2000;7(20):1744–52. PubMed PMID: 11083496.

[57] Alam T, Sollinger HW. Glucose-regulated insulin production in hepatocytes. Transplantation. 2002;74(12):1781–7. PubMed PMID: 12499898.

[58] Kuwahata M, Kuramoto Y, Sawai Y, Amano S, Tomoe Y, Segawa H, et al. Polypyrimidine tract-binding protein is involved in regulation of albumin synthesis in response to food intake. Journal of Nutritional Science and Vitaminology. 2008;54(2):142–7. PubMed PMID: WOS:000255558500005.

[59] Groskreutz DJ, Sliwkowski MX, Gorman CM. Genetically-engineered proinsulin constitutively processed and secreted as mature, active insulin. Journal of Biological Chemistry. 1994;269(8):6241–5. PubMed PMID: WOS:A1994MY84000112.

[60] Yanagita M, Hoshino H, Nakayama K, Takeuchi T. Processing of mutated proinsulin with tetrabasic cleavage sites to mature insulin reflects the expression of furin in nonendocrine cell-lines. Endocrinology. 1993;133(2):639–44. PubMed PMID: WOS:A1993LQ84600031.

[61] Kaarsholm NC, Norris K, Jorgensen RJ, Mikkelsen J, Ludvigsen S, Olsen OH, et al. Engineering stability of the insulin monomer fold with application to structure-activity relationships. Biochemistry. 1993;32(40):10773–8. PubMed PMID: 8399225.

[62] Cantore A, Ranzani M, Bartholomae CC, Volpin M, Valle PD, Sanvito F, et al. Liver-directed lentiviral gene therapy in a dog model of hemophilia B. Science Translational Medicine. 2015;7(277):277ra28. PubMed PMID: 25739762.

[63] Duckworth WC, Bennett RG, Hamel FG. Insulin degradation: progress and potential. Endocrine Reviews. 1998;19(5):608–24. PubMed PMID: WOS:000076368400006.

[64] Alam T, Wai P, Held D, Vakili ST, Forsberg E, Sollinger H. Correction of diabetic hyperglycemia and amelioration of metabolic anomalies by minicircle DNA mediated glucose-dependent hepatic insulin production. PLoS One. 2013;8(6):e67515. PubMed PMID: 23826312. Pubmed Central PMCID: 3694888.

[65] Morgan RA, Anderson WF. Human gene-therapy. Annual Review of Biochemistry. 1993;62:191–217. PubMed PMID: WOS:A1993LK68700008.

[66] Volpers C, Kochanek S. Adenoviral vectors for gene transfer and therapy. Journal of Gene Medicine. 2004;6:S164–71. PubMed PMID: WOS:000220114400015.

[67] Short DK, Okada S, Yamauchi K, Pessin JE. Adenovirus-mediated transfer of a modified human proinsulin gene reverses hyperglycemia in diabetic mice. American Journal of Physiology – Endocrinology and Metabolism. 1998;275(5):E748–56. PubMed PMID: WOS:000076913700004.

[68] Parks RJ, Chen L, Anton M, Sankar U, Rudnicki MA, Graham FL. A helper-dependent adenovirus vector system: removal of helper virus by Cre-mediated excision of the viral packaging signal. Proceedings of the National Academy of Sciences of the United States of America. 1996;93(24):13565–70. PubMed PMID: 8942974. Pubmed Central PMCID: 19345.

[69] Parks RJ, Graham FL. A helper-dependent system for adenovirus vector production helps define a lower limit for efficient DNA packaging. Journal of Virology. 1997;71(4): 3293–8. PubMed PMID: 9060698. Pubmed Central PMCID: 191467.

[70] Marshall E. Clinical trials—Gene therapy death prompts review of adenovirus vector. Science. 1999;286(5448):2244–5. PubMed PMID: WOS:000084318500002.

[71] Assessment of adenoviral vector safety and toxicity: report of the National Institutes of Health Recombinant DNA Advisory Committee. Human Gene Therapy. 2002;13(1): 3–13. PubMed PMID: WOS:000173075000002.

[72] Hausl MA, Zhang W, Muther N, Rauschhuber C, Franck HG, Merricks EP, et al. Hyperactive sleeping beauty transposase enables persistent phenotypic correction in mice and a canine model for hemophilia B. Molecular Therapy: The Journal of the American Society of Gene Therapy. 2010;18(11):1896–906. PubMed PMID: 20717103. Pubmed Central PMCID: 2990515.

[73] Muzyczka N. Use of adenoassociated virus as a general transduction vector for mammalian-cells. Current Topics in Microbiology. 1992;158:97–129. PubMed PMID: WOS:A1992JB19700006.

[74] Park YM, Woo SA, Lee GT, Ko JY, Lee Y, Zhao ZS, et al. Safety and efficacy of adeno-associated viral vector-mediated insulin gene transfer via portal vein to the livers of streptozotocin-induced diabetic Sprague–Dawley rats. Journal of Gene Medicine. 2005;7(5):621–9. PubMed PMID: WOS:000229297100010.

[75] Muzzin P, Eisensmith RC, Copeland KC, Woo SLC. Hepatic insulin gene expression as treatment for type 1 diabetes mellitus in rats. Molecular Endocrinology. 1997;11(6):833–7. PubMed PMID: WOS:A1997XA52800020.

[76] Cavazzana-Calvo M, Hacein-Bey S, Basile CD, Gross F, Yvon E, Nusbaum P, et al. Gene therapy of human severe combined immunodeficiency (SCID)-X1 disease. Science. 2000;288(5466):669–72. PubMed PMID: WOS:000086727100043.

[77] Ronen K, Negre O, Roth S, Colomb C, Malani N, Denaro M, et al. Distribution of lentiviral vector integration sites in mice following therapeutic gene transfer to treat beta-thalassemia. Molecular Therapy. 2011;19(7):1273–86. PubMed PMID: WOS: 000292295900014.

[78] Papayannakos C, Daniel R. Understanding lentiviral vector chromatin targeting: working to reduce insertional mutagenic potential for gene therapy. Gene Therapy. 2013;20(6):581–8. PubMed PMID: WOS:000319976600001.

[79] Schweizer M, Merten OW. Large-scale production means for the manufacturing of lentiviral vectors. Current Gene Therapy. 2010;10(6):474–86. PubMed PMID: WOS: 000285778400006.

[80] Kafri T, Blomer U, Peterson DA, Gage FH, Verma IM. Sustained expression of genes delivered directly into liver and muscle by lentiviral vectors. Nature Genetics. 1997;17(3):314–7. PubMed PMID: 9354796.

[81] Hwang BY, Schaffer DV. Engineering a serum-resistant and thermostable vesicular stomatitis virus G glycoprotein for pseudotyping retroviral and lentiviral vectors. Gene Therapy. 2013;20(8):807–15. PubMed PMID: 23364315. Pubmed Central PMCID: 3735647.

[82] Ren B, O'Brien BA, Swan MA, Koina ME, Nassif N, Wei MQ, et al. Long-term correction of diabetes in rats after lentiviral hepatic insulin gene therapy. Diabetologia. 2007;50(9): 1910–20. PubMed PMID: WOS:000248771800016.

[83] Ren BH, O'Brien BA, Byrne MR, Ch'ng E, Gatt PN, Swan MA, et al. Long-term reversal of diabetes in non-obese diabetic mice by liver-directed gene therapy. Journal of Gene Medicine. 2013;15(1):28–41. PubMed PMID: WOS:000314178500005.

Chapter 7

Application of Genome Editing Technology to MicroRNA Research in Mammalians

Lei Yu, Jennifer Batara and Biao Lu

Additional information is available at the end of the chapter

http://dx.doi.org/10.5772/

Abstract

Targeted nucleases have recently emerged as a powerful genome editing tool. The ability of introducing targeted, desired changes into mammalian genome makes them an invaluable tool to unravel functions of miRNAs in biology and disease. In combination with homologous donor vector, targeted nucleases can achieve high efficiency and precision, enabling bi-allelic ablation of miRNA in cultured somatic cells. Here we review the structure and function of miRNA as well as the unique implementation of genome editing technology in modifying miRNA sequences in mammalians. This chapter discusses the four mainstay genome editing technologies: meganuclease, zinc finger nuclease (ZFN), transcription activator-like effector nuclease (TALEN) and clustered regularly interspaced short palindromic repeat-associated nuclease Cas9 (CRISPR-Cas9), focusing on TALEN.

Keywords: Genome editing, miroRNA, TALEN, ZFN, CRISPR-Cas9, Meganuclease, Non-homologous end joining, Homologous recombination

1. Introduction

Recent achievements in genome editing have resulted in progress beyond any imagination decades ago. This new technology provides tools for fast and precise alterations into genome with unprecedented efficiency and specificity. For instance, a group of targeted nucleases have been successfully used to modify genomic sequences in a wide variety of cells and organisms, including mammalians.[1-4] This modification is realized by the combination of conventional gene knockout technology with genome editing, empowering the researchers to make any desired changes in a given gene or its regulatory elements to establish casual linkage between genetic variations and biological phenotypes. This new methodology completes a framework for studying gene function and regulation *in vitro* and *in vivo*.

1.1. Conventional gene knockout technology

Mammalian genomes contain billions of DNA bases and are difficulty to manipulate.[5-7] Conventional genome editing is inefficient and must undergo a number of complicated and time-consuming procedures to obtain double knockouts. Because this technology had been built upon homologous recombination (HR), the targeted gene modifications occur at an extremely low frequency (1 in 10^6 ~ 10^9 cells).[8] This is the reason why this conventional approach is mainly used for producing knockout mice, for which a large amount of high quality embryonic stem cells is available. Although knockout mice have provided great insights into the fundamentals of developmental biology and physiology; the pathological roles of genes have yet to be addressed in relevant disease models or human samples.

1.2. Novel genome editing technology: Targeted nucleases

To overcome limitations of conventional gene knockout technology, a number of nuclease-based methods have been developed in the past decade.[9-12] These novel technologies exploit the ability of endonucleases to induce double-stranded DNA breaks (DSBs) and stimulate subsequent damage repair mechanisms in mammalian cells.[11, 13-17] Remarkably, this nuclease-induced DSBs not only make sequence changes at the break points but also enhance HR rate to an astonish frequency of 1-40%.[7, 18] This approach provides a simple and effective way to streamline the genome editing with the potential to generate double knockout models in both cell lines and animals. Because of this technical breakthrough, nuclease-based methods quickly gain popularity for gene editing in variety of cell types and species. To date, four major classes of targeted nucleases are prominent: meganucleases derived from microbial mobile elements, zinc finger nuclease (ZFN) based on eukaryotic transcription factor DNA binding motif, transcription activator-like effector nuclease (TALEN) derived from a plan-invasive bacterial protein and most recently the bacterial type II clustered, regularly interspaced, short palindromic repeat-associated nuclease Cas (CRISPRE-Cas).[19] These technologies are collectively termed as targeted nucleases because they all have more or less a programmable DNA binding domain for specific genome targeting. The first three nucleases all recognize specific DNA sequences through protein-DNA interaction, whereas the last CRISPRE-Cas is targeted by a short guide RNA that recognizes the unwind DNA via Watson-Crick base pairing. The purpose of the nuclease is to make localized DSBs. These DSBs will invoke powerful non-homologous end-joining as well as homologous-direct recombination repair pathway for the versatile and precise modification of complicated mammalian genome.[14, 18, 20-32]

2. miRNA

Large fraction of the genome is transcribed without protein coding potential. [33, 34] These non-coding transcripts can be broadly categorized into short and long non-coding RNAs, in which the arbitrary size delineation is at ~ 200 bases in length. Short non-coding RNAs are less 200 bases including microRNA, tRNAs and snoRNAs. An abundant class of small regulatory

RNAs are termed microRNAs (miRNAs) that have been identified to play major roles in gene regulation.[35-38] Like coding mRNAs, miRNAs are transcribed by polymerase II and processed into ~ 22 nucleotide non-coding transcripts, which repress translation by binding to target sites within the 3′ untranslated region of mRNA.[39-41] Recent genomic studies have discovered over 1000 miRNAs in human and mouse genome.[34] It is estimated that up to 80% of human genes are regulated by miRNAs.[36, 42] Some mRNAs are clustered in polycistronic transcripts and allow coordinated regulation, while others are expressed in a tissue-specific and developmental stage-specific manner.[35, 40] The roles of miRNAs have been extensively studied at molecular and cellular levels in a number of species including mammals.[43, 44] Furthermore, their roles in physiology and development in animals have been established by conditional knockins and knockouts.[45-47] In fact, miRNAs are expressed across genome, and many of them show spatial and temporal expression. Similar to transcription factors, miRNAs may function as master regulators of embryonic pluripotency, differentiation, and tissue/organ formation.[37] Recently, increasing evidence suggests that miRNAs are implicated in numerous disease states as well. For instance, miR-21 was found to be overexpressed in virtually all types of human cancers and thus has emerged as an important therapeutic target in cancer treatment.[45, 48, 49] Additionally, miR-21 and other miRNAs have been shown to play crucial regulatory roles in basic cell functions such as cell growth, proliferation and apoptosis.[48, 49] Their pathological roles in multiple human diseases such as autoimmune, cardiovascular and neurological disorders and obesity are emerging.[50, 51]

2.1. Genome organization and biogenesis

MicroRNA genes have a great diversity in mammalian genomes. They are located in either introns of annotated protein-coding genes or outside the context of an annotated genes.[34] Gemone analysis studies reveal that up to 42% of human miRNAs are in clusters of two or more with pairwise chromosome distance of at most 3000 nucleotides. This pattern of clustering allows similar levels of expression and coordinated regulation. Examples of the two most famous clusters are miR-17~92 and miR-302~367. While the deletion of miR-17~92 cluster causes skeletal and growth defects, overexpression of miR-302-367 cluster enhances somatic reprogramming to pluripotent status. These clustered miRNAs appear to function together; therefore alteration of a specific member may not result in expected changes in physiology.

Although the exact mechanism of the regulation of each miRNA remains to be determined, the biogenesis of miRNA is becoming more apparent. From their gene loci, miRANs are initially transcribed by RNA polymerase II as long primary transcripts, which are processed into approximately 70-nucleotide precursors by the RNAse III enzyme Drosha in the nucleus.[52] These highly structured precursors are termed pre-miRNAs and subsequently transported from the nucleus to the cytoplasm by an Exportin-5 protein shuttler.[40] In cytoplasm, these pre-miRNAs are further cleaved by another RNase III enzyme, Dicer, resulting in imperfect miRNA:miRNA duplexes of about 18 ~ 24 bp in length.[53, 54] Although either strand of the duplex may potentially mature as a functional miRNA, only one strand is usually chosen and subsequently incorporated into the RNA-induced silence complex (RISC) where the miRNA and its mRNA target interact.[55, 56]

2.2. Molecular tools for miRNA study

Despite their fundamental importance, the exact role of miRNAs in the context of human development and disease processes remain largely unknown. This is partly due to lack of effective methods for completely abolishing their expression in human cells and disease-relevant models. Although miRNA knockdown by short-interfering RNAs (siRNAs) provides a rapid and inexpensive tool to study most protein-coding genes, it cannot be used to reduce mature miRNAs in a sensible way at the cellular level. Other alternatives include small molecule inhibitors, antisense oligo-nucleotides, anti-miR vectors and miRNA sponges.[57, 58] However, the major limitations of these methods are transient nature of their effects and a high risk of off-target effects as well as toxicity. It is no surprise to see reports on discrepancies between the effects of miRNA inhibitors and genetic knockouts. With the completion of the genome sequencing, genome-wide gene targeting knockout of miRNA have been taken.[59] As a result, resources for the conditional ablation of miRNAs in mouse have been attempted. [60] In one study, generation of 162 miRNA targeting vectors and 46 germline-transmitted miRNA knockout mice was reported.[60] However, this homologous donor based knockout technology is expected to have tremendous hurdles to apply to other mammalian species such as rat, pig and monkeys. These larger animals are more close to humans; however, they all lack large scale culture of high quality embryonic cells to make use of this conventional technology. Nevertheless, the work in mice will provide an important basis for elucidating the physiological roles of certain miRNAs in at least one animal species.

3. Targeted nucleases

Programmable DNA-binding proteins have emerged as an exciting platform for engineering targeted nucleases for precise genome editing.[61] The key component of these engineered nucleases is the DNA recognition domain that is capable of directing the nucleases to the specific genome loci therefore generating DNA double-strand breaks (DSBs) near or at the target sites.[62] In mammalian cells, these DSBs are repaired by one of two mechanisms, non-homologous end-joining (NHEJ) or homologous recombination (HR).[18] Repair by NHEJ is error-prone and often results in small insertions or deletion mutations, termed indels, at the break point. At the same time, DSBs can also greatly stimulate high-fidelity HDR repair mechanism in fast dividing cells. In mammalian cells, NHEJ is dominant over HR, but the frequency of latter can be substantially increased when large amount of exogenous homologous donors are co-delivered into cells.[25]

With these strategies, various methods, including meganucleases, ZFNs, TALENs and CRISPRE-Cas have been reported for genome editing in a wide variety of mammalian species. [63-70] For these methods, a DNA endonuclease enzyme for generating DSBs is brought in place by a guide molecule. In the first three scenarios, the guider is a protein, whereas the last (CRESPRE-Cas) is a short stretch of RNA. Figure 1 illustrates the action of targeted nucleases and the binding mode of guide molecules. In general, the protein guider is more specific than the RNA guider, where degeneration is governed by hybridization mechanism. The pros and cons as well as their features of four main technologies are discussed below.

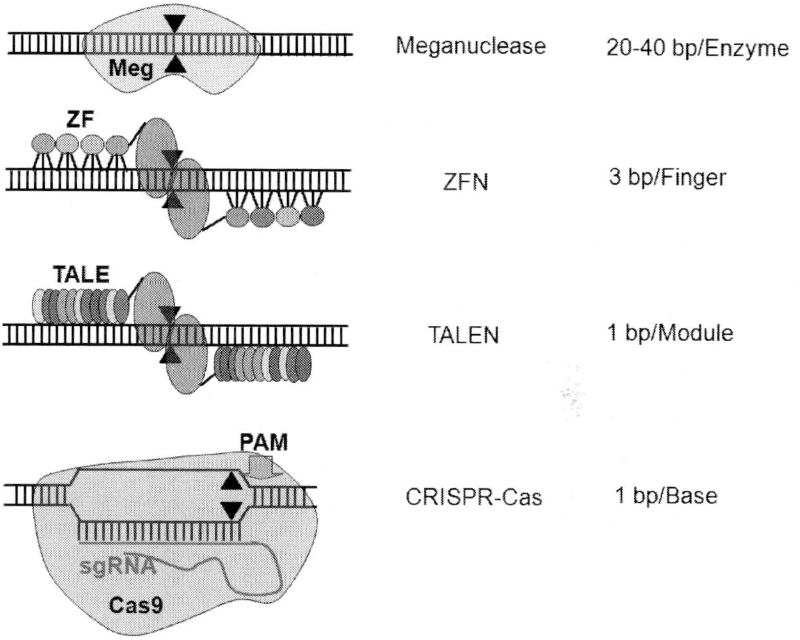

Figure 1. Action of the Targeted Nucleases and Their DNA Binding Mode. Genomic DNA is shown horizontally in black and double stranded, with the site of DNA cleavage indicated by arrowheads. For meganuclease, the holoenzyme binds and cleaves the target DNA. For ZFN and TALENs, they function as a pair; with one zinc finger DNA binding domain (ZF) binding to the upper strand while the other ZF binding to the lower strand. Once this fused FokI enzyme (purple oval) is oriented to form a homodimer, it is activated to cut DNA. For CRISPR-Cas, the Cas9 holoenzyme (orange oval) is directed to the target site by the guide RNA and cleave the DNA at the position close to the PAM motif (grey arrow).

3.1. Meganuclease

Meganucleases are endonucleases with a large DNA recognition site of 12 ~ 45 bp in length.[71] As a result, this site may occur only once in most mammalian genome.[6] Although Meganuclease possesses high degree precision and low toxicity, its target range is limited. Moreover, the intertwined DNA-binding domain and nuclease domain restrict its capacity to reprogram for other targets, and the probability of finding a meganuclease for cutting a desired locus is extremely slim.[72] For example, the phiC31 integrase mediates recombination between a donor DNA of two 34 bp sequences, termed as attachment sites (att) and the other in mammalian genome. In the introduction of phage integrase, a phiC31 integrase can insert a plasmid donor DNA of any size and requires no additional co-factors. Other advantages of using phiC31 integrase include non-viral delivery, sustainable transgene expression, and functions in species like bacteria, yeast, plants, frogs, chickens, mice, rats, pigs, cows, and humans.[2, 73] However, potential sites of phiC31-based genome editing are limited. In human genome, of the 106 mapped integration sites of phiC31, ~ 39% are within coding genes and ~ 61% are in the intergenic regions.[74]

3.2. Zinc finger nuclease

First described in 1996, ZFN is a chimeric protein that is composed of two distinctive domains, a programmable DNA binding domain and an endonuclease, FokI.[75] Because of its programmability, ZFN have been successfully employed to modify almost all genome types including bacteria, plants, and animals.[1, 76, 77] In fact, ZFNs have been used for the correction of a number of hereditary diseases such as hemophilia B, sickle cell anaemia, a-1 antitrypsin deficiency and gene therapies for viral infections.[77-80] ZFN-based HIV gene therapy is already under clinical trials, because of its specificity and safety profile.[81, 82] However, the use of ZFNs has been partly hindered by the complex and time-consuming strategies to generate highly specific zinc-finger arrays with sufficient affinity and specificity. It is worthy to note that ZFNs also suffer some constrains in targeting range with about one potential target site every 500 bp.[83] It is conceivable that one may find it difficult to design ZFNs to precisely targeting a smaller gene in genome, such as miRNAs.

3.3. Transcription activator-like effector nuclease

Similar to ZFNs, TALENs are also fusion proteins that comprise of a programmable DNA binding domain with an endonuclease, FokI. The DNA-binding domain of TALEN consists of ~34 amino acid repeats, followed by a single half repeat of 20 amino acids.[84] Interestingly, the tandem repeats are nearly identical, except for two amino acid codons at positions 12 and 13, referred to as "repeat variable di-residue" (RVD). [85]Each of the four most common RVDs specifies the binding to one of the four nucleotide bases (Table 1). The natural RVD for G is NN with asparagine at positions 12 and 13. NN binds G with high affinity, but also recognizes and binds A with relative low affinity.[86] Although artificial NH or NK provides good specificity for G, the binding affinity to G is relative low as compared to NN.[86] It is worth to note that TALE proteins bind DNA sequences with an invariable base T in the first position of the target. The corresponding module is not the repeat but the cryptic sequences flanking the repeats. Because the first binding base is invariable, the DAN-binding domain can theatrically be programmed to bind any sequences starting with a "T". Taking advantage of the simplicity of the coding principle, the DNA-binding domain can be easily designed to allow binding of almost any sequences within genomes. Owing to the repetitive nature of the DNA binding domain, the assembly of the custom TALENs by direct synthesis or traditional cloning is expensive and technically challenging.[87] Realizing the potential of TALEN technology, a number of approaches for TALEN assembly have been devised to allow low ~ medium throughput, or high-throughput with automation. Fortunately, a number of Biotech companies provide either assembly kit and/or service due to its technical difficulty (Table 2). Like ZFN, this genome editing technology has been shown to function in a wide variety of cells and organisms, including bacteria, yeast, plant, insect, zebrafish and mammal.[21, 88-92] Furthermore, unlike meganuclease or ZFN that limit the choice of targets, TALEN can virtually bind any loci in the genome with new design that removes the 5′ first "T" base constrain.

34 amino-acid repeat	RVD	Deoxynucleotide
LTPEQVVAIASNIGGKQALETVQRLLPVLCQAHG	NI	A
LTPEQVVAIASNGGGKQALETVQRLLPVLCQAHG	NG	T
LTPEQVVAIASNNGGKQALETVQRLLPVLCQAHG	NN(NH*/NK*)	G or A
LTPEQVVAIASHDGGKQALETVQRLLPVLCQAHG	HD	A

*Note: NH and NK favor specificity rather than activity. They bind to G more specifically but with less affinity as compared to NN.

Table 1. TALEN DNA binding repeat and its simple code scheme

Company	Web Address	Tool Kit and Service
Addgene (Cambridge, Massachusetts)	www.addgene.org	Collections of TALEN assembly kits for multiple organisms
Cellectis (Paris, France)	www.cellectis.com	TALEN or meganuclease-based services
GenCopoea (Rockville, Maryland)	www.genecopoeia.com	TALEN design and assembly; HR donor design and cloning; Indel detection kits
GenScript (Piscataway, New Jersey)	www.genscript.com	TALEN gene synthesis
Life Technologies (Carlsbad, California)	www.lifetechnologies.com	TALEN design and assembly services for genome editing; Targeted gene activator/ silencer cloning services

Table 2. Selected companies with TALEN tool kit and service

3.4. CISPRE-Cas9

Unlike ZFNs or TALENs, CRISPR-Cas needs a short RNA for target site recognition, which is mediated by Watson-Crick DNA base pairing. To form a triplex with genomic DNA, CRISPR-Cas9 also requires a NGG protospacer adjacent motif (PAM) immediately downstream of the hybrid region.

The bacterial CRISPRE-Cas9 system can be reconstituted in mammalian cells using three components: a programmable, specificity-determining CRISPR RNA (crRNA), an auxiliary trans-activating (tracrRNA), and a CRISPRE-associated endonuclease Cas9 (Cas9).[93, 94] In current applications, crRNA and tracrRAN duplexes are fused to generate a chimeric sgRAN that mimics the natural crRNA-tracrRNA hybrid. The single sgRNA has been shown to interact with the holoenzyme Cas9 to generate efficient cleavage.[95] The adaptation of the CRISPR-Cas9 by the research community have been phenomenal, and the system has been used to modify genes in most model species, including drosophila, silkworm, plant, C elegan, zebrafish, Xenopus, rat, mouse, pig, and human.[96-106] The success is mainly due to some significant advantages. First, a single protein Cas9 is required and remains the same. This means no time-consuming protein engineering is required, in contrast to that of meganuclease,

ZFN or TALEN. Second, genome targeting depends on an oligonucleotides (20 ~ 30 in length), which are very easy and cheap to produce. Third, among the established programmable DNA-binding domains, CRISPR-Cas9 is most easily to facilitate genome-scale perturbations owing to its one-binding to one target. Last, this is an open system, where most of them are established and can be purchased from the non-profit distributor Addgene in Cambridge Massachusetts. However, the sequence of individual guide RNA makes difference in terms of efficacy and specificity, not all guide RNA provides the same high levels of genome editing activities.[107] Bad guide RAN can elicit off-target effects, causing cyto-toxicity to cells.[108] This can be very problematic for therapeutic applications.

The technologies are still quickly evolving, and tools to better assess both efficiency and specificity will be essential to improve the system. Choosing certain method shall be carefully balanced to serve the need of research. For example, ZFNs, TALENs and CRISPR-Cas9 are reasonable options to target most coding genes. ZFNs and TALENs shall be preferred when specificity and low-toxicity are required. For genome-wide screen or in dealing with multiple targets, CRISCR-Cas9 is more suitable because of its robustness and one molecule for one target. For small gene manipulations, TALENs may be of choice since they have broad rang access in the genome and edit precisely the target with high specificity with much less toxicity and off-target effects as compared to CRISCR-Cas9.

4. Design strategy and experimental approaches

4.1. Design strategies for miRNA targeting

One of the challenges in knocking out a miRNA is that the mature and fully functional miRNA is only ~ 22 nucleotides in length. Therefore, sequence alterations outside this 22-nucleotide region may have little or no effects on function of miRNA. To design useful targeted nucleases, sites have to be carefully chosen so that targeted nucleases are directed to the critical region of miRNA gene loci. To this end, the relative small but highly structured pre-miRNA (~70 bp) appears appropriate since it contains sequences vital for miRNA biogenesis and function.[35, 43] Within pre-miRNA, the seed regions of 5p and 3p as well as the dicer processing sites are of choice (Figure 2). Indeed, a number of reports have used this strategy for miRNA knockouts; studies have shown that small indels targeting these sites are effective to abolish miRNA expression in mammalian cells.[45, 46, 59]

Here, we use miR-21 as an example to illustrate the valid target sites that were successfully used in knockout studies. We choose miR-21 for a number of reasons. First, the human miRNA-21 was one of the first mammalian genes identified. MicroRNA-21 is located on plus strand of chromosome within a protein coding gene TMEM49. It is independently transcribed as a ~3433 nuclotides long primary transcript, where the pre-miR-21 (72 nucleotides) has a typical stem-loop (hairpin) structure similar to other miRNAs.[109] Second, the mature miRNA sequence is typically processed from the 5′-arm of the miR-21 precursor.[110] Last, miR-21 is strongly conserved and its role in physiology and pathology has been extensively explored and firmly established.[45, 48, 49, 109] Together, miR-21 provides general design strategies for miRNA gene editing.

As illustrated in Figure 2, the region corresponding to 1-8 nucleotides of the mature miRNA is the most preferred site of cleavage. This is because a small indel in this seed region is expected to abolish miRNA activity.[111] The second choice would be the adjacent Dicer processing site of the seed region. In principle, the chosen of process sites shall be preferred to the one that close to the seed region, indels involves both process site and seed sequences would most likely results in complete knockout. According to above consideration, the 5′ miRNA seed region and the adjacent Dicer processing site are usually chosen for miR-21 knockout. Similarly, the 3′-arm of the miRNA precursor shall be preferred when the mature miRNA is of 3p miRNA.

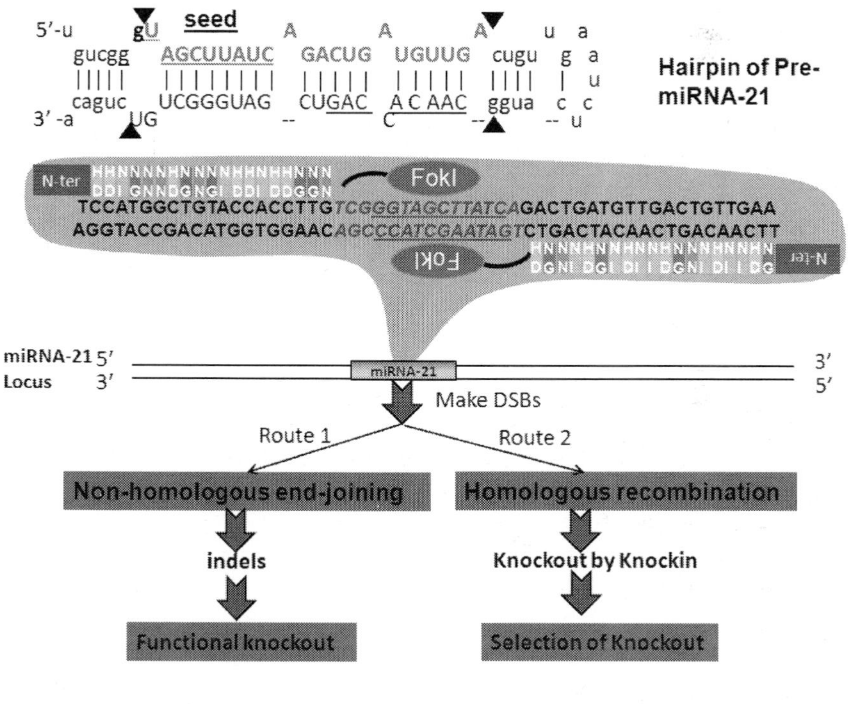

Figure 2. MicroRNA knockout strategy and the preferred cleavage sites. The upper panel shows the stem-loop structure of miR-21, with mature miR-21 shown in red and seed region underlined. The middle panel shows the TALEN pair and their binding sequences, separated by a 15-bp spacer. The cleavage site normally falls in the middle of the spacer, where the FokI dimerizes and makes double-strand breaks (DSBs) at the seed and/or Dicer process site. These DSBs can lead to two potential consequences as illustrated in the low panel. The predominant route 1 leads to indels via non-homologous end joining mechanism, where the alternative route 2 may lead to a precise genome editing via homologous recombination. Repair by homologous recombination can be used to bring in any desired changes at the targeted site. For knocking out miRNA-21, replacement of pre-miRNA with selection markers can facilitate the enrichment and selection of edited cells.

Here we introduce the potential sites as a general strategy for miRNA knockout utilizing the dominant NHEJ mechanisms. In fact ZFNs, TALENs and CRISPR-Cas have been used to functionally knockout a number of miRNAs in mammalian cells.[45, 59, 112-116] Among the three types of programmable nucleases, TALENs are suitable for small loci with narrow targetable regions, whereas ZFNs and CRISPR-Cas are limited by availability of binding modules and the requirement of PAM motif respectively. It is important to keep in mind; the binding sites of targeted nucleases may be different from the cleavage sites in terms of design. For ZFNs and TALENs, the cleavage site is situated in the middle of their binding sites of ZFNs or TALENs. For CRSPR-Cas, the cleavage site is within the guider RNA and close to the PAM motif (Figure 2).

4.2. TALEN design and assembly

Free online tools such as TALEN Targeter and SAPTA are available to design DNA binding domain of TALEN that is specific and has a low risk of of-target effects.[117, 118] Similar to ZFNs, TALENS s requires the design of protein pairs, which bind two optimal anchoring positions on opposite strands, usually spaced 15 ~ 25 bp apart to allow for FokI dimerization and cutting.[119] Length of DNA binding sequences may vary, typically ranging from 14 ~ 20 bp. In humans, 20 bp may offer high specificity, considering the genome size. It is worthy to note that longer domains (18 ~ 20) may decrease cell toxicity by reducing the risk of off-target effects.[120]

A number of approaches have been developed for rapid assembly of custom TALENs. With these advances, TALEN pairs can be generated easily and economically in a matter of days. [121] Most of the methods rely on the ability of type IIS restriction enzymes to assemble premade repeats into fully functional TALEN scaffold (Figure 3). A TALEN scaffold is comprised of a number of domains from the N-terminus to the C-terminus, including a nuclear localization signal, part of the N-terminal sequences for the first "T" recognition, flowed by the last half-repeat and part of C-terminal sequences fused with a nuclease FokI. Because the last binding nucleotide can be any one of the four, the TALEN scaffolds normally have four different flavours. To make the TALEN fully functional, an eukaryotic gene promoter is normally placed at 5′ of TALEN coding region and a poly A signal sequences at the 3′ end of TALEN.

Most assembly protocol is based on the Golden Gate method, which relies on the ability of type IIS restriction enzymes to cut outside their recognition site. Type IIS recognition sites arranged in inverse orientation at the 5′ and 3′ end of DNA fragment will be removed upon cleavage, slowing simultaneous restriction and ligation. The continuous re-digestion of unwanted ligation products increases the formation of the desired construct.[122, 123] As type IIS fusion sites can be designed to have different sequences, Golden Gate cloning enables directional and seamless assembly of multiple DNA fragments. Based on this principle, one-step or two-step assembly protocol or kits are developed and commercially available to allow do-it-yourself assembly of TALEN in any molecular biology laboratories.[121]

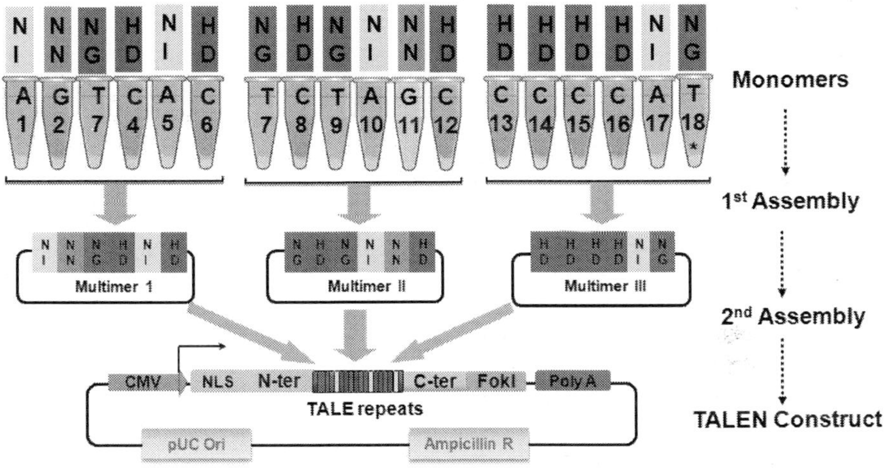

Figure 3. Golden Gate assembly of TALENs and molecular structure. Two-step assembly strategy can increase the correct joining of monomers into multimers and subsequently into the TALEN scaffold. A TALEN, from its N- to C-terminus, is composed of a nuclear localization signal peptide (NLS), modified N- and C-termini flanking the assembled RVDs, and fused nuclease FokI at C-terminus. CMV (cytomegaloviral) promoter situates at the 5′ of the TALEN and drives its expression in mammalian cells. The polyadenylation signal (Poly A) is to add a poly(A) tail to the TALEN mRNA.

4.3. HR donor design and construction

While it is possible to disrupt genes in mammalian genome with TALENs alone, the frequency of gene editing is typically 2 ~ 40%, averaging ~16% for mono-allelic disruptions.[87] Cells carrying bi-allelic disruption are rare and require time-consuming signal cell-derivation and subsequent screening.[5] One strategy is to combine TALENs targeting to the miRNA seed region with a homologous recombination of donor vector carrying a selectable marker.[46, 59, 114] This approach enables convenient positive selection, and the combination of NHEJ with stem-loop deletion results in efficient bi-allelic miRNA gene ablation, which can be as high as >90%.[46]Additionally, by using HR donors, endogenous loci can be potentially modified with custom sequences such as IRES-florescent proteins to allow functional assessment of endogenous miRNA expression and regulation.[124]

4.4. Generation of knockout cell clones

All targeted nucleases can be used in mammalians to create miRNA knockouts.[45, 46, 59, 113-116] Success of miRNA gene editing depends largely on the ability to deliver all the reagents efficiently to the cells. For transgenics, direct injection of DNA vector or sometime *in vitro* transcribed nuclease mRNAs into embryos is effective.[88, 125] For cultured cells, the options include plasmid DNA transfection, viral deliver, and transfection with synthetic mRNA or proteins.[126-130] Normally, donor DNAs can also supplied via plasmid format and co-delivered with nucleases using the same methodology. To deal with difficult cell types, viral transduction appears to be effective. Although lentiviral system appears to work well for both ZFNs and CRISPR-Cas, it is not suitable for TALEN delivery due to its incompatiblility with the tandem repeats of TALEN.

Following the delivery, small-scale sequence changes are introduced at the break by NHEJ.[18] These indels are typically assayed by polymerase chain reaction (PCR) amplification of the region, followed by DNA sequencing or by a gel electrophoresis assay using T7E1 endonuclease, or alternatively by high-resolution melting analysis.[131] In addition to detecting changes at genomic level, a reverse transcripts-PCR on the expression can also be performed to confirm the reduced/ablated expression of miRNA at transcript level. PCR-based genotyping can be used to make distinctions between HR and NHEJ events when donor DNA is used.[131]

5. Applications of targeted nucleases for miRNA research

Targeted nucleases are powerful genome editing tools for uncovering gene functions. Though relatively new, they have been successfully employed in a broad variety of systems and produced exciting results for miRNA research.

5.1. Knockout of the miR-200 family

To understand the biological and pathological significance of miRNAs, a talen-based knockout library for 274 highly conserved human miRNAs has been established.[59] To demonstrate the genome editing activities of the TALEN library, 66 TALEN pairs against 33 miRNA loci are selected. All TALEN pairs tested induce mutations as assessed by a mismatch-senstivie T7EI assay with a frequency above 0.5%.[132] To gain some insight of functional role miRNAs, the authors conduct detail analysis on members of the miR-200 family.

There are at least two members in the highly conserved miR-200 family, miR-141 and miR-200c.[59] Interestingly, miR-141 and miR-200c have largely indistinguishable activity and differ only in the seed region by one nucleotide. This imposes a great difficulty to use either over-expression or complementary inhibitor-based knockdown to investigate the potential functional divergence without complications. Using TALEN technology, however, the authors can target the seed region of the 5p strand for miR-141 but choose the Drosha processing site in the 3p strand for miR-200c, hence avoid cross-targeting.[52, 110] With this design, both single and double knockout clones were obtained and their corresponding expression of mature miRNAs was confirmed by RT-PCR.[59] Using these cell models, the authors found that miR-141 represses the expression of mRNAs that have miR-141 motif at the 3′-untranslated region. Similarly, miR-200c represses expression of mRNAs that have miR-200c motif. These data indicate the two closely related miRNAs do not cross-react notably and may control largely nonoverlapping group of genes. Together, TALEN-based method may provide unprecedented tools for miRNA research with great precision and specificity.

5.2. miR-21 ablation and cancer research

MicroRNA-21 gene knockout in the cultured human cells was achieved independently by two research groups.[45, 131, 133] The first group used a combination of TALEN pair with a HR donor. TALENs were designed to position the miR-21 seed region in the central portion of the spacer, directing the cleavage to the functionally essential miRNA motif. The HR donor construct was created corresponding to the cleavage location of the TALEN pair and carried 509-bp (5′ arm) and 600-bp (3′ arm) regions of homology to the miR-21 genome sequences. TALEN pair and HR donor were delivered together using transfection protocol. In the case of HR events, the donor replaces the entire miR-21 precursor with two selectable marker genes (red florescent protein and puromycin resistant gene). Clonal population of cells in which an HR event occurred can be easily selected by puromycin treatment. Because NJEJ is the predominant repair mechanism induced by DSBs, selecting for HR events would most likely produce clones that harbour bi-allelic modifications, with the second allele carrying an indel in the seed region. In fact, this approach demonstrated bi-allelic miR-21 gene disruption at very high frequency of 87% in cultured HEK293 cells.[131] Analysis of three independent clones showed a total loss of miR-21expression. Phenotypical examination revealed an increase in miR-21 target gene expression, reduced cell proliferation, and alteration of global miRNA expression profiles, which is in agreement with the role of miR-21 in cancer biology.[45, 48]

5.3. miRNA knockout for transgenics

To explore the gene specific-function *in vivo*, generation of gene-targeting animal is powerful strategy. To this end, targeted nucleases have demonstrated that all nucleases are applicable for generating knockout animal and have several advantages over the conventional embryonic stem cells or nuclear transfer technologies. Recently, ZFNs, TALEN and CRISPRE-Cas were able to generate knockout animals in several model species in addition to the established spices of mice.[19] One of the strong merit is that they are free of constrain from using embryonic stem cells. The genome of the fertilized zygotes can be directly modified by injecting the targeted nuclease. Additionally, these new tools are robust enough to allow a simultaneous knockout of multiple genes, thus greatly accelerating the experimental speed and answering complicated biological questions.

Targeted knockout of miRNAs in mice by TALEN has also been reported.[113] In this study, microinjection of synthesized mRNA of TALEN was carried out in one cell stage of embryo. Embryos were allowed to develop to two-cell stage and subsequently transferred into pseudopregnant female mice. With optimized protocol, these mice were able to produce 29.6% mono-allelic offspring. Further cross-mating of the heterozygous founder mice with the wild type strain was able to produced heterozygote offspring, suggesting transmittable of the mutated miRNA.

6. Prospects and challenges

Targeted nuclease technology has become one the most powerful and versatile platforms for engineering biology. The new technology is enabling systematic interrogation of genome, including miRNAs in mammals with high precision and efficacy. It is superb over other technologies by creating null mutations that lead to complete suppression of gene expression. Furthermore, bi-allelic ablation of miRNAs in cultured somatic cells opens new avenue to study the pathological role of miRNAs in relevant disease samples. Knowledge gained from these studies is more likely lead to discovery of novel drug targets, accelerating new therapeutics towards clinic application. In deed, a number of animal studies as well as clinical trials using targeted nucleases already provide encouraging results. In addition to repairing mutations underlying inherited diseases, targeted nucleases can be used to create productive mutations in tissues to combat viral infections or complex diseases such as familial hypercholesterolemia and hypertension.

Targeted nucleases can be tweaked to carry out other functions, such as modifying DNA associated histones, activating or inhibiting gene transcriptions and monitoring chromatin dynamics in living cells. It becomes increasing apparent that targeted nuclease is a versatile and common platform for elucidating gene function and epigenetic regulation.

However, many challenges still lie ahead. The most prominent issue may be the off-target effect. In this aspect, ZFNs and TALENs appear less problematic owing to the intimate interaction with their binding sites and the requirement of correct binding of two molecules. In contrast, CRISPRE-Cas tends to have higher off-target effects, because it binds DNA via base pairing. One or more mispairings may be tolerated, which may cause detrimental unintended effects.[134] Attempts to solve this issue include improvement of guide RNA scaffold, use of computational algorithms to predict off-targets, and development of high throughput method to assess unintended cuts. The challenge, though, is difficult to detect the off-target cleavage. Hence, one shall interpret the data with caution, and is aware of the potential off-target effects and its related toxicity. Secondly, although nuclease molecule may have high level of activity within cells, one may find that some tissue types are still difficult to deliver into. In this case, multiple delivery methods shall be tried, suitable one shall be optimized. Finally, for further development toward clinical application, it will be essential to thoroughly the safety and toxicity profile using a variety of mammalian models.

7. Conclusion

Advance in genome engineering technology based on the targeted nucleases are enabling the systematic interrogation of mammalian gene function. Using this system, miRNA gene sequences within mammalian genome can be easily edited with high efficacy and precision. Targeted miRNA editing will empower researcher to reveal the complex regulatory circuits governed by miRNAs and to realize, in the long-term, their full diagnostic and therapeutic potential.

http://dx.doi.org/10.5772/

Author details

Lei Yu[1], Jennifer Batara[2] and Biao Lu[2*]

*Address all correspondence to: blu2@scu.edu

1 Institute for Advanced Interdisciplinary Research in Science and Technology, East China Normal University, Shanghai, China

2 Department of Bioengineering, Santa Clara University, Santa Clara, California, USA

References

[1] Petolino, J.F., *Genome editing in plants via designed zinc finger nucleases.* In Vitro Cell Dev Biol Plant, 2015. 51(1): p. 1-8.

[2] Petersen, B. and H. Niemann, *Molecular scissors and their application in genetically modified farm animals.* Transgenic Res, 2015. 24(3): p. 381-96.

[3] Kandavelou, K., et al., *Targeted manipulation of mammalian genomes using designed zinc finger nucleases.* Biochem Biophys Res Commun, 2009. 388(1): p. 56-61.

[4] Wefers, B., et al., *Generation of targeted mouse mutants by embryo microinjection of TALEN mRNA.* Nat Protoc, 2013. 8(12): p. 2355-79.

[5] Vasquez, K.M., et al., *Manipulating the mammalian genome by homologous recombination.* Proc Natl Acad Sci U S A, 2001. 98(15): p. 8403-10.

[6] Carninci, P., et al., *The transcriptional landscape of the mammalian genome.* Science, 2005. 309(5740): p. 1559-63.

[7] Bollag, R.J., A. Watdman, and R.M. Liskay, *Homologous recombination in mammalian cells.* Annu Rev Genet, 1989. 23(1): p. 199-225.

[8] Segal, D.J. and J.F. Meckler, *Genome engineering at the dawn of the golden age.* Annu Rev Genomics Hum Genet, 2013. 14: p. 135-58.

[9] Lloyd, A., et al., *Targeted mutagenesis using zinc-finger nucleases in Arabidopsis.* Proc Natl Acad Sci U S A, 2005. 102(6): p. 2232-7.

[10] Durai, S., et al., *Zinc finger nucleases: custom-designed molecular scissors for genome engineering of plant and mammalian cells.* Nucleic Acids Res, 2005. 33(18): p. 5978-90.

[11] Christian, M., et al., *Targeting DNA double-strand breaks with TAL effector nucleases.* Genetics, 2010. 186(2): p. 757-61.

[12] Wu, Y., et al., *Correction of a genetic disease in mouse via use of CRISPR-Cas9.* Cell Stem Cell, 2013. 13(6): p. 659-62.

[13] Ran, F.A., et al., *Double nicking by RNA-guided CRISPR Cas9 for enhanced genome editing specificity.* Cell, 2013. 154(6): p. 1380-9.

[14] Morton, J., et al., *Induction and repair of zinc-finger nuclease-targeted double-strand breaks in Caenorhabditis elegans somatic cells.* Proc Natl Acad Sci U S A, 2006. 103(44): p. 16370-5.

[15] Fung, H. and D.M. Weinstock, *Repair at single targeted DNA double-strand breaks in pluripotent and differentiated human cells.* PLoS One, 2011. 6(5): p. e20514.

[16] Takasu, Y., et al., *The use of TALENs for nonhomologous end joining mutagenesis in silkworm and fruitfly.* Methods, 2014. 69(1): p. 46-57.

[17] Do, T.U., et al., *Zinc Finger Nuclease induced DNA double stranded breaks and rearrangements in MLL.* Mutat Res, 2012. 740(1-2): p. 34-42.

[18] Lieber, M.R., *The mechanism of double-strand DNA break repair by the nonhomologous DNA end-joining pathway.* Annu Rev Biochem, 2010. 79: p. 181-211.

[19] Gaj, T., C.A. Gersbach, and C.F. Barbas, 3rd, *ZFN, TALEN, and CRISPR/Cas-based methods for genome engineering.* Trends Biotechnol, 2013. 31(7): p. 397-405.

[20] Bachu, R., I. Bergareche, and L.A. Chasin, *CRISPR-Cas targeted plasmid integration into mammalian cells via non-homologous end joining.* Biotechnol Bioeng, 2015.

[21] Budhagatapalli, N., et al., *Targeted Modification of Gene Function Exploiting Homology-Directed Repair of TALEN-Mediated Double Strand Breaks in Barley.* G3 (Bethesda), 2015.

[22] Lee, C.M., et al., *Correction of the DeltaF508 Mutation in the Cystic Fibrosis Transmembrane Conductance Regulator Gene by Zinc-Finger Nuclease Homology-Directed Repair.* Biores Open Access, 2012. 1(3): p. 99-108.

[23] Mahfouz, M.M., et al., *De novo-engineered transcription activator-like effector (TALE) hybrid nuclease with novel DNA binding specificity creates double-strand breaks.* Proc Natl Acad Sci U S A, 2011. 108(6): p. 2623-8.

[24] Maresca, M., et al., *Obligate ligation-gated recombination (ObLiGaRe): custom-designed nuclease-mediated targeted integration through nonhomologous end joining.* Genome Res, 2013. 23(3): p. 539-46.

[25] Maruyama, T., et al., *Increasing the efficiency of precise genome editing with CRISPR-Cas9 by inhibition of nonhomologous end joining.* Nat Biotechnol, 2015. 33(5): p. 538-42.

[26] Nakade, S., et al., *Microhomology-mediated end-joining-dependent integration of donor DNA in cells and animals using TALENs and CRISPR/Cas9.* Nat Commun, 2014. 5: p. 5560.

[27] Overlack, N., et al., *Gene repair of an Usher syndrome causing mutation by zinc-finger nuclease mediated homologous recombination.* Invest Ophthalmol Vis Sci, 2012. 53(7): p. 4140-6.

[28] Ponce de Leon, V., et al., *Generation of TALEN-mediated GRdim knock-in rats by homologous recombination.* PLoS One, 2014. 9(2): p. e88146.

[29] Vannocci, T., et al., *Nuclease-stimulated homologous recombination at the human beta-globin gene.* J Gene Med, 2014. 16(1-2): p. 1-10.

[30] Weinthal, D.M., R.A. Taylor, and T. Tzfira, *Nonhomologous end joining-mediated gene replacement in plant cells.* Plant Physiol, 2013. 162(1): p. 390-400.

[31] Xue, H., et al., *Genetic Modification in Human Pluripotent Stem Cells by Homologous Recombination and CRISPR/Cas9 System.* Methods Mol Biol, 2014.

[32] Yuan, J., et al., *A zinc finger nuclease assay to rapidly quantitate homologous recombination proficiency in human cell lines.* Anal Biochem, 2013. 434(1): p. 96-8.

[33] Alexander, R.P., et al., *Annotating non-coding regions of the genome.* Nat Rev Genet, 2010. 11(8): p. 559-71.

[34] Esteller, M., *Non-coding RNAs in human disease.* Nat Rev Genet, 2011. 12(12): p. 861-74.

[35] Bartel, D.P., *MicroRNAs: genomics, biogenesis, mechanism, and function.* Cell, 2004. 116(2): p. 281-297.

[36] Bartel, D.P., *MicroRNAs: target recognition and regulatory functions.* Cell, 2009. 136(2): p. 215-233.

[37] Marson, A., et al., *Connecting microRNA genes to the core transcriptional regulatory circuitry of embryonic stem cells.* Cell, 2008. 134(3): p. 521-533.

[38] He, L. and G.J. Hannon, *MicroRNAs: small RNAs with a big role in gene regulation.* Nature reviews genetics, 2004. 5(7): p. 522-531.

[39] Lee, Y., et al., *MicroRNA genes are transcribed by RNA polymerase II.* The EMBO journal, 2004. 23(20): p. 4051-4060.

[40] Lee, Y., et al., *MicroRNA maturation: stepwise processing and subcellular localization.* The EMBO journal, 2002. 21(17): p. 4663-4670.

[41] Kim, Y.K. and V.N. Kim, *Processing of intronic microRNAs.* The EMBO journal, 2007. 26(3): p. 775-783.

[42] Hobert, O., *Gene regulation by transcription factors and microRNAs.* Science, 2008. 319(5871): p. 1785-1786.

[43] Chen, X., *MicroRNA biogenesis and function in plants.* FEBS Lett, 2005. 579(26): p. 5923-5931.

[44] Grimm, D., et al., *Fatality in mice due to oversaturation of cellular microRNA/short hairpin RNA pathways.* Nature, 2006. 441(7092): p. 537-541.

[45] Chen, B., et al., *Disruption of microRNA-21 by TALEN leads to diminished cell transformation and increased expression of cell-environment interaction genes.* Cancer Lett, 2015. 356(2 Pt B): p. 506-16.

[46] Uhde-Stone, C., et al., *A TALEN-based strategy for efficient bi-allelic miRNA ablation in human cells.* RNA, 2014. 20(6): p. 948-55.

[47] Thai, T.-H., et al., *Regulation of the germinal center response by microRNA-155.* Science, 2007. 316(5824): p. 604-608.

[48] Asangani, I., et al., *MicroRNA-21 (miR-21) post-transcriptionally downregulates tumor suppressor Pdcd4 and stimulates invasion, intravasation and metastasis in colorectal cancer.* Oncogene, 2008. 27(15): p. 2128-2136.

[49] Si, M., et al., *miR-21-mediated tumor growth.* Oncogene, 2007. 26(19): p. 2799-2803.

[50] Pauley, K.M., S. Cha, and E.K. Chan, *MicroRNA in autoimmunity and autoimmune diseases.* Journal of autoimmunity, 2009. 32(3): p. 189-194.

[51] Takanabe, R., et al., *Up-regulated expression of microRNA-143 in association with obesity in adipose tissue of mice fed high-fat diet.* Biochem Biophys Res Commun, 2008. 376(4): p. 728-732.

[52] Ballarino, M., et al., *Coupled RNA processing and transcription of intergenic primary microRNAs.* Mol Cell Biol, 2009. 29(20): p. 5632-8.

[53] Lee, Y.S., et al., *Distinct roles for Drosophila Dicer-1 and Dicer-2 in the siRNA/miRNA silencing pathways.* Cell, 2004. 117(1): p. 69-81.

[54] Chendrimada, T.P., et al., *TRBP recruits the Dicer complex to Ago2 for microRNA processing and gene silencing.* Nature, 2005. 436(7051): p. 740-744.

[55] Dykxhoorn, D.M., C.D. Novina, and P.A. Sharp, *Killing the messenger: short RNAs that silence gene expression.* Nature Reviews Molecular Cell Biology, 2003. 4(6): p. 457-467.

[56] McManus, M.T. and P.A. Sharp, *Gene silencing in mammals by small interfering RNAs.* Nature reviews genetics, 2002. 3(10): p. 737-747.

[57] Krützfeldt, J., et al., *Silencing of microRNAs in vivo with 'antagomirs'.* Nature, 2005. 438(7068): p. 685-689.

[58] Ebert, M.S., J.R. Neilson, and P.A. Sharp, *MicroRNA sponges: competitive inhibitors of small RNAs in mammalian cells.* Nat Methods, 2007. 4(9): p. 721-726.

[59] Kim, Y.K., et al., *TALEN-based knockout library for human microRNAs.* Nat Struct Mol Biol, 2013. 20(12): p. 1458-64.

[60] Park, C.Y., et al., *A resource for the conditional ablation of microRNAs in the mouse.* Cell Rep, 2012. 1(4): p. 385-391.

[61] Fichtner, F., R. Urrea Castellanos, and B. Ulker, *Precision genetic modifications: a new era in molecular biology and crop improvement.* Planta, 2014. 239(4): p. 921-39.

[62] Aiba, Y., J. Sumaoka, and M. Komiyama, *Artificial DNA cutters for DNA manipulation and genome engineering.* Chem Soc Rev, 2011. 40(12): p. 5657-68.

[63] Beumer, K., et al., *Efficient gene targeting in Drosophila with zinc-finger nucleases.* Genetics, 2006. 172(4): p. 2391-403.

[64] Geurts, A.M., et al., *Knockout Rats via Embryo Microinjection of Zinc-Finger Nucleases.* Science, 2009. 325(5939): p. 433.

[65] Galli, C., et al., *Genetic engineering including superseding microinjection: new ways to make GM pigs.* Xenotransplantation, 2010. 17(6): p. 397-410.

[66] Carroll, D., K.J. Beumer, and J.K. Trautman, *High-efficiency gene targeting in Drosophila with zinc finger nucleases.* Methods Mol Biol, 2010. 649: p. 271-80.

[67] Yu, S., et al., *Highly efficient modification of beta-lactoglobulin (BLG) gene via zinc-finger nucleases in cattle.* Cell Res, 2011. 21(11): p. 1638-40.

[68] Barzel, A., et al., *Native homing endonucleases can target conserved genes in humans and in animal models.* Nucleic Acids Res, 2011. 39(15): p. 6646-59.

[69] Petersen, B. and H. Niemann, *Advances in genetic modification of farm animals using zinc-finger nucleases (ZFN).* Chromosome Res, 2015. 23(1): p. 7-15.

[70] Zhou, X., et al., *Generation of CRISPR/Cas9-mediated gene-targeted pigs via somatic cell nuclear transfer.* Cell Mol Life Sci, 2015. 72(6): p. 1175-84.

[71] Silva, G., et al., *Meganucleases and other tools for targeted genome engineering: perspectives and challenges for gene therapy.* Curr Gene Ther, 2011. 11(1): p. 11.

[72] Carroll, D., *Using nucleases to stimulate homologous recombination.* Methods Mol Biol, 2004. 262: p. 195-207.

[73] Teimourian, S. and R. Abdollahzadeh, *Technology developments in biological tools for targeted genome surgery.* Biotechnol Lett, 2015. 37(1): p. 29-39.

[74] Chalberg, T.W., et al., *Integration specificity of phage ϕC31 integrase in the human genome.* J Mol Biol, 2006. 357(1): p. 28-48.

[75] Kim, Y.G., J. Cha, and S. Chandrasegaran, *Hybrid restriction enzymes: zinc finger fusions to Fok I cleavage domain.* Proc Natl Acad Sci U S A, 1996. 93(3): p. 1156-60.

[76] de Pater, S., et al., *ZFN-induced mutagenesis and gene-targeting in Arabidopsis through Agrobacterium-mediated floral dip transformation.* Plant Biotechnol J, 2009. 7(8): p. 821-35.

[77] Qu, X., et al., *Zinc finger nuclease: a new approach for excising HIV-1 proviral DNA from infected human T cells.* Mol Biol Rep, 2014. 41(9): p. 5819-27.

[78] Li, H., et al., *In vivo genome editing restores haemostasis in a mouse model of haemophilia.* Nature, 2011. 475(7355): p. 217-21.

[79] Wayengera, M., *Zinc finger nucleases for targeted mutagenesis and repair of the sickle-cell disease mutation: An in-silico study.* BMC Blood Disord, 2012. 12: p. 5.

[80] Yusa, K., et al., *Targeted gene correction of alpha1-antitrypsin deficiency in induced pluripotent stem cells.* Nature, 2011. 478(7369): p. 391-4.

[81] Tebas, P., et al., *Gene editing of CCR5 in autologous CD4 T cells of persons infected with HIV.* N Engl J Med, 2014. 370(10): p. 901-10.

[82] Jo, Y.I., H. Kim, and S. Ramakrishna, *Recent developments and clinical studies utilizing engineered zinc finger nuclease technology.* Cell Mol Life Sci, 2015.

[83] Jabalameli, H.R., et al., *Zinc finger nuclease technology: advances and obstacles in modelling and treating genetic disorders.* Gene, 2015. 558(1): p. 1-5.

[84] Zhang, M., et al., *TALE: a tale of genome editing.* Prog Biophys Mol Biol, 2014. 114(1): p. 25-32.

[85] Jankele, R. and P. Svoboda, *TAL effectors: tools for DNA targeting.* Brief Funct Genomics, 2014. 13(5): p. 409-19.

[86] Christian, M.L., et al., *Targeting G with TAL effectors: a comparison of activities of TALENs constructed with NN and NK repeat variable di-residues.* PLoS One, 2012. 7(9): p. e45383.

[87] Wright, D.A., et al., *TALEN-mediated genome editing: prospects and perspectives.* Biochem J, 2014. 462(1): p. 15-24.

[88] Nakajima, K. and Y. Yaoita, *Highly efficient gene knockout by injection of TALEN mRNAs into oocytes and host transfer in Xenopus laevis.* Biol Open, 2015. 4(2): p. 180-5.

[89] Zhang, Z., et al., *Disruption of Zebrafish Follicle-stimulating Hormone Receptor (fshr) but Not Luteinizing Hormone Receptor (lhcgr) Gene by TALEN Leads to Failed Follicle Activation in Females Followed by Sexual Reversal to Males.* Endocrinology, 2015: p. en20151039.

[90] Biffi, A., *Clinical translation of TALENS: Treating SCID-X1 by gene editing in iPSCs.* Cell Stem Cell, 2015. 16(4): p. 348-9.

[91] Valton, J., et al., *Efficient strategies for TALEN-mediated genome editing in mammalian cell lines.* Methods, 2014. 69(2): p. 151-70.

[92] Niu, Q., et al., *Knockout of the adp gene related with colonization in Bacillus nematocida B16 using customized transcription activator-like effectors nucleases.* Microb Biotechnol, 2015. 8(4): p. 681-92.

[93] Doench, J.G., et al., *Rational design of highly active sgRNAs for CRISPR-Cas9-mediated gene inactivation.* Nat Biotechnol, 2014. 32(12): p. 1262-7.

[94] Liu, L. and X.D. Fan, *CRISPR-Cas system: a powerful tool for genome engineering.* Plant Mol Biol, 2014. 85(3): p. 209-18.

[95] Jinek, M., et al., *A programmable dual-RNA-guided DNA endonuclease in adaptive bacterial immunity.* Science, 2012. 337(6096): p. 816-21.

[96] Hwang, W.Y., et al., *Efficient genome editing in zebrafish using a CRISPR-Cas system.* Nat Biotechnol, 2013. 31(3): p. 227-9.

[97] Friedland, A.E., et al., *Heritable genome editing in C. elegans via a CRISPR-Cas9 system.* Nat Methods, 2013. 10(8): p. 741-3.

[98] Horii, T., et al., *Genome engineering of mammalian haploid embryonic stem cells using the Cas9/RNA system.* PeerJ, 2013. 1: p. e230.

[99] Mashiko, D., et al., *Generation of mutant mice by pronuclear injection of circular plasmid expressing Cas9 and single guided RNA.* Sci Rep, 2013. 3: p. 3355.

[100] Ren, X., et al., *Optimized gene editing technology for Drosophila melanogaster using germ line-specific Cas9.* Proc Natl Acad Sci U S A, 2013. 110(47): p. 19012-7.

[101] Frokjaer-Jensen, C., *Exciting prospects for precise engineering of Caenorhabditis elegans genomes with CRISPR/Cas9.* Genetics, 2013. 195(3): p. 635-42.

[102] Upadhyay, S.K., et al., *RNA-guided genome editing for target gene mutations in wheat.* G3 (Bethesda), 2013. 3(12): p. 2233-8.

[103] Xie, F., et al., *Seamless gene correction of beta-thalassemia mutations in patient-specific iPSCs using CRISPR/Cas9 and piggyBac.* Genome Res, 2014. 24(9): p. 1526-33.

[104] Nakayama, T., et al., *Simple and efficient CRISPR/Cas9-mediated targeted mutagenesis in Xenopus tropicalis.* Genesis, 2013. 51(12): p. 835-43.

[105] Whitworth, K.M., et al., *Use of the CRISPR/Cas9 system to produce genetically engineered pigs from in vitro-derived oocytes and embryos.* Biol Reprod, 2014. 91(3): p. 78.

[106] Shao, Y., et al., *CRISPR/Cas-mediated genome editing in the rat via direct injection of one-cell embryos.* Nat Protoc, 2014. 9(10): p. 2493-512.

[107] Wiles, M.V., et al., *CRISPR-Cas9-mediated genome editing and guide RNA design.* Mamm Genome, 2015.

[108] Semaan, M., D. Ivanusic, and J. Denner, *Cytotoxic Effects during Knock Out of Multiple Porcine Endogenous Retrovirus (PERV) Sequences in the Pig Genome by Zinc Finger Nucleases (ZFN).* PLoS One, 2015. 10(4): p. e0122059.

[109] Selcuklu, S.D., M.A. Donoghue, and C. Spillane, *miR-21 as a key regulator of oncogenic processes.* Biochem Soc Trans, 2009. 37(4): p. 918.

[110] Lee, Y., et al., *The nuclear RNase III Drosha initiates microRNA processing.* Nature, 2003. 425(6956): p. 415-419.

[111] Lewis, B.P., C.B. Burge, and D.P. Bartel, *Conserved seed pairing, often flanked by adenosines, indicates that thousands of human genes are microRNA targets.* Cell, 2005. 120(1): p. 15-20.

[112] Saydaminova, K., et al., *Efficient genome editing in hematopoietic stem cells with helper-dependent Ad5/35 vectors expressing site-specific endonucleases under microRNA regulation.* Mol Ther Methods Clin Dev, 2015. 1: p. 14057.

[113] Takada, S., et al., *Targeted gene deletion of miRNAs in mice by TALEN system.* PLoS One, 2013. 8(10): p. e76004.

[114] Hu, R., et al., *Targeting human microRNA genes using engineered Tal-effector nucleases (TALENs).* PLoS One, 2013. 8(5): p. e63074.

[115] Ho, T.T., et al., *Targeting non-coding RNAs with the CRISPR/Cas9 system in human cell lines.* Nucleic Acids Res, 2015. 43(3): p. e17.

[116] Zhang, Z., et al., *Dissecting the roles of miR-302/367 cluster in cellular reprogramming using TALE-based repressor and TALEN.* Stem Cell Reports, 2013. 1(3): p. 218-25.

[117] Lin, Y., et al., *SAPTA: a new design tool for improving TALE nuclease activity.* Nucleic Acids Res, 2014. 42(6): p. e47.

[118] Doyle, E.L., et al., *TAL Effector-Nucleotide Targeter (TALE-NT) 2.0: tools for TAL effector design and target prediction.* Nucleic Acids Res, 2012. 40(W1): p. W117-W122.

[119] Cermak, T., et al., *Efficient design and assembly of custom TALEN and other TAL effector-based constructs for DNA targeting.* Nucleic Acids Res, 2011. 39(12): p. e82.

[120] Reyon, D., et al., *FLASH assembly of TALENs for high-throughput genome editing.* Nat Biotechnol, 2012. 30(5): p. 460-5.

[121] Uhde-Stone, C., et al., *A do-it-yourself protocol for simple transcription activator-like effector assembly.* Biol Proced Online, 2013. 15(1): p. 3.

[122] Sanjana, N.E., et al., *A transcription activator-like effector toolbox for genome engineering.* Nat Protoc, 2012. 7(1): p. 171-92.

[123] Cermak, T., C.G. Starker, and D.F. Voytas, *Efficient design and assembly of custom TALENs using the Golden Gate platform.* Methods Mol Biol, 2015. 1239: p. 133-59.

[124] Scherr, M., et al., *Lentivirus-mediated antagomir expression for specific inhibition of miRNA function.* Nucleic Acids Res, 2007. 35(22): p. e149-e149.

[125] Schwartz, H.T. and P.W. Sternberg, *Transgene-free genome editing by germline injection of CRISPR/Cas RNA.* Methods Enzymol, 2014. 546: p. 441-57.

[126] Holkers, M., T. Cathomen, and M.A. Goncalves, *Construction and characterization of adenoviral vectors for the delivery of TALENs into human cells.* Methods, 2014. 69(2): p. 179-87.

[127] Senis, E., et al., *CRISPR/Cas9-mediated genome engineering: an adeno-associated viral (AAV) vector toolbox.* Biotechnol J, 2014. 9(11): p. 1402-12.

[128] Jia, J., et al., *Bacterial delivery of TALEN proteins for human genome editing.* PLoS One, 2014. 9(3): p. e91547.

[129] Mock, U., et al., *mRNA transfection of a novel TAL effector nuclease (TALEN) facilitates efficient knockout of HIV co-receptor CCR5.* Nucleic Acids Res, 2015. 43(11): p. 5560-71.

[130] Liu, C., et al., *Generation of outbred Ace2 knockout mice by RNA transfection of TALENs displaying colitis reminiscent pathophysiology and inflammation.* Transgenic Res, 2015. 24(3): p. 433-46.

[131] Uhde-Stone, C., et al., *A TALEN-based strategy for efficient bi-allelic miRNA ablation in human cells.* RNA, 2014.

[132] Cho, S.W., et al., *Targeted genome engineering in human cells with the Cas9 RNA-guided endonuclease.* Nat Biotechnol, 2013. 31(3): p. 230-2.

[133] Ma, X., et al., *Loss of the miR-21 allele elevates the expression of its target genes and reduces tumorigenesis.* Proceedings of the National Academy of Sciences, 2011. 108(25): p. 10144-10149.

[134] Pellagatti, A., et al., *Application of CRISPR/Cas9 genome editing to the study and treatment of disease.* Arch Toxicol, 2015. 89(7): p. 1023-34.

Chapter 8

Emerging Gene Correction Strategies for Muscular Dystrophies: Scientific Progress and Regulatory Impact

Houria Bachtarzi and Tim Farries

Additional information is available at the end of the chapter

http://dx.doi.org/10.5772/62282

Abstract

Muscular dystrophies comprise a heterogeneous cluster of inherited muscle degenerative disorders with the common feature of progressive muscle weakness. These represent good candidates for treatment with gene-based therapies. Progress in gene transfer technologies has raised hopes for successful therapeutic restoration of mutated genes such as dystrophin in Duchenne muscular dystrophy. Delivery to enough muscle cells, however, remains a challenge for a successful gene replacement therapy. Other approaches based on exon skipping to correct mutant dystrophin's pre-mRNA splicing patterns have been tried, and partial restoration of dystrophin expression was reported in late-stage clinical trials, but full therapeutic efficacy is yet to be confirmed. The emergence of gene editing and its recent success in AIDS have opened a new therapeutic era for muscular dystrophies. This chapter will cover new gene correction strategies for muscular dystrophies and their regulatory challenges before they can become routine treatment modalities in the clinic.

Keywords: gene therapy, muscular dystrophies, regulatory impact, inherited muscle degenerative disorders, progressive muscle weakness

1. Introduction

Muscular dystrophy refers to a range of conditions of progressive muscle weakening generally due to genetic defects in proteins that are critical for muscle functioning. The most common form of the disease, and one of the more common of any seriously debilitating genetic disorders, is Duchenne muscular dystrophy (DMD), which is caused by the mutation of the dystrophin gene on the X chromosome and found in approximately 1 in 3600 of male births. The dystrophin gene product is a large protein with repeated elements and is involved in connecting the muscle fibres to the extracellular matrix [1]. Other proteins, mutations of which may cause forms of muscular dystrophy, include poly(A) binding protein nuclear 1 (PABPN1), myotonic dystrophy (DM) protein kinase (DMPK), or the product of the Emery-Dreifuss muscular dystrophy gene.

This chapter focuses on opportunities to develop treatments for these conditions with therapies that target the genetic defect that is the cause of the disease. The promise shown by new technologies for gene therapeutics makes this a particularly propitious time to consider the impact that such scientific progress may have on this clinically important set of conditions.

Notable among the new technological advances of relevance are the development of methods of specifically changing the sequence of the human genome, either by introducing a new gene sequence, which may repair the function of a defective gene, or by correcting the defect in the endogenous gene. Although the means to deliver genes to human cells still rely on the use of viral vectors, the methodology for the effective manufacture of these complex biologicals has developed. The approvals by the European Union (EU) of Glybera® and Imlygic® show that viral products can be manufactured to the quality standards acceptable for commercial products, with acceptable profiles of safety and efficacy. Another transformative technology is the derivation from stem cell precursors of different differentiated cell types with the potential to repair damaged or otherwise defective tissue. Putting these technologies together results in a possibility to create and manufacture to GMP standard, well-characterised patient-specific (or at least patient-compatible), genetically modified cells, a combination that may be anticipated in the near future to enable the viable therapeutic treatment of many previously intractable genetic diseases.

This chapter looks at the current status and future prospects of how the latest gene-based technologies are being applied to the alleviation, or even cure, of inherited muscular dystrophies. Furthermore, we consider the challenge of translating these treatment modalities into medicines that can be approved for commercial use by the regulatory authorities, notably within the EU. To do this, developers must go beyond the scientific mechanisms of efficacy to establish how the products will be manufactured to consistent quality standards and to demonstrate that they are clinically safe.

2. Muscular dystrophies

Muscular dystrophies comprise a heterogeneous cluster of inherited muscle degenerative disorders, each caused by a distinct gene mutation. More than 30 genes have been identified, each causing a different type of muscle pathology with different patterns of muscle weakness and disease progression (**Table 1**). This includes but is not limited to Duchenne, Becker, congenital, myotonic, Emery-Dreifuss, facioscapulohumeral, oculopharyngeal, and limb-girdle muscular dystrophies. Each of these varies in terms of pattern of inheritance, age of disease onset, biochemical markers (such as creatinine kinase's upper limits), types of muscles affected, and complications at other organ sites, including cardiac and pulmonary problems. These give rise to varied symptoms, including muscle weakness and wasting (a common feature of a number of muscular dystrophies), joint stiffness, and scoliosis in addition to respiratory complications (such as chest infections and shortness of breath). Other symptoms include ankle swelling often linked with cardiomyopathy, fainting, eyelid drooping, and dysphagia. Ophthalmological symptoms, such as myopia in facioscapulohumeral muscular dystrophy and severe congenital muscular dystrophy (CMD) variants, cataracts in DM, and eyelid drooping in oculopharyngeal muscular dystrophy (OPMD), have also been reported. Hearing loss and skin lesions are also common in facioscapulohumeral muscular dystrophy and Ullrich CMD, respectively. Overall, these cause profound impairments in physical activity and quality of life [2, 3].

Disease type	Affected protein	Genetics	Age of onset (years)	Main pathological features	Current treatment
		Duchenne muscular dystrophy (DMD) and Becker muscular dystrophy (BMD)			
DMD	Dystrophin	Xp21.2 (X-R)	<5	Progressive muscle weakness (proximal to distal); cardiomyopathy; respiratory impairment.	No cure available to date. Symptomatic treatment to control the onset of symptoms and maximise the quality of life (corticosteroid therapy, physiotherapy, orthopaedic assistance, respiratory support).
BMD	Dystrophin	Xp21.2 (X-R)	5–40	Milder disease phenotype compared to DMD; variable progression; cardiomyopathy.	No cure available to date. Symptomatic treatment to control symptoms and improve the quality of life.
		Limb girdle muscular dystrophy (LGMD)			
LGMD type 1B	Lamin A/C	1q21.2 (AD)	10–30	Variable symptoms.	No cure available to date.
LGMD type 1C	Caveolin-3	3p25 (AD)		Progressive muscle weakness and atrophy; cardiomyopathy and respiratory impairment observed in sarcoglycanopathies (LGMD	Symptomatic care to control cardiac and respiratory symptoms with medications and/or devices; physical
LGMD type 2A	Calpain-3	15q15.1 (AR)			

Disease type	Affected protein	Geneti cs	Age of onset (years)	Main pathological features	Current treatment
LGMD type 2B	Dysferlin	2p13 (AR)		types 2C, 2D, 2E, and 2F), dystroglycanopathies (LGMD types 2I, 2K, 2M, and 2N), and LGMD type 2L.	activity and orthopaedic assistance.
LGMD type 2L	Anoctamin 5	11p14.3 (AR)			
Sarcoglycanopathies					
LGMD type 2C	γ-Sarcoglycan	13q12 (AR)	10–30	Progressive muscle weakness and atrophy; cardiomyopathy and respiratory impairment observed in sarcoglycanopathies (LGMD types 2C, 2D, 2E, and 2F), dystroglycanopathies (LGMD types 2I, 2K, 2M, and 2N), and LGMD type 2L.	Symptomatic care to control cardiac and respiratory symptoms with medications and/or devices; physical activity and orthopaedic assistance.
LGMD type 2D	α-Sarcoglycan	17q12-q21.33 (AR)			
LGMD type 2E	β-Sarcoglycan	4q12 (AR)			
LGMD type 2F	δ-Sarcoglycan	5q33 (AR)			
Dystroglycanopathies					
LGMD type 2I	Fukutin-related protein	19q13.3 (AR)	10–30		
LGMD type 2K	Protein-1-O-mannosyl-transferase 1	9q34 (AR)			
LGMD type 2M	Fukutin	9q31 (AR)			
LGMD type 2N	Protein-O-mannosyl-transferase 2	14q24 (AR)			
		Congenital muscular dystrophy (CMD)			
MDC1A	Laminin α2 chain of merosin	6q2 (AR)	Congenital onset	Progressive muscle weakness; severity of disease progression varies; respiratory impairment observed in CMD with merosin deficiency, WWS, MDC1C, Ullrich syndrome.	No cure available to date. No specific treatment existent. Current pharmaceutical management includes a combination of orthopaedic supportive care, regular monitoring, and intervention to treat and
MDC1C	Fukutin-related protein	19q13 (AR)			
MDC1D	Like-glycosyl transferase	22q12 (AR)			
FCMD	Fukutin	9q31-q33 (AR)			

http://dx.doi.org/10.5772/62282

Disease type	Affected protein	Genetics	Age of onset (years)	Main pathological features	Current treatment
Fukutin-defective WWS	Fukutin	9q31-q33 (AR)			control respiratory insufficiency.
Protein-O-mannosyl-transferase 1 defective WWS	Protein-1-O-mannosyl-transferase 1	9q34 (AR)			
Protein-O-mannosyl-transferase 2 defective WWS	Protein-O-mannosyl-transferase 2	14q24 (AR)			
Protein-O-linked mannose β1,2-N-aminyltransferase 1 defective WWS	Protein-O-linked mannose β1,2-*N*-aminyltransferase 1	1p34 (AR)			
Fukutin-related protein defective WWS	Fukutin-related protein	19q13 (AR)			
Ullrich syndrome-collagen type VI subunit $\alpha1$ defect	Collagen type VI, subunit $\alpha1$	21q22.3 (AR)			
Ullrich syndrome-collagen type VI subunit $\alpha2$ defect	Collagen type VI, subunit $\alpha2$	21q22.3 (AR)			
Ullrich syndrome-collagen type VI subunit $\alpha3$ defect	Collagen type VI, subunit $\alpha3$	2q37 (AR)			
Integrin $\alpha7$-defective CMD	Integrin $\alpha7$	12q13 (AR)			
Integrin $\alpha9$-defective CMD	Integrin $\alpha9$	3p21.3 (AR)			
Facioscapulohumeral muscular dystrophy	Unknown	4q35 (AD), 18 (AD)	7–30	Slowly progressive disease with facial and shoulder muscle weaknesses; uncommon/mild cardiac impairment and infrequent respiratory problems. Hearing loss reported.	No cure available to date. Symptomatic treatment and physiotherapy.

Disease type	Affected protein	Genetics	Age of onset (years)	Main pathological features	Current treatment
Emery-Dreifuss	Emerin	Xq28 (XLR, AD)	4–20	Slow-progressive muscle weakness and wasting affecting upper arms, lower legs, shoulder, and hips; cardiac problems (arrhythmia, conduction block); infrequent respiratory problems.	No cure available to date. Current management aims to prevent sudden cardiac death (use of pacemakers, cardiac transplantation) and correction of skeletal complications to maintain ambulation (Achilles tenotomy, surgery, orthopaedic support).
OPMD	PABPN1	14q11.2 (AD, AR)	30–60	Slow-progressive muscle weakness affecting muscles of the eyes (ptosis), throat (dysphagia), and proximal limb. Aspiration pneumonia reported. No cardiac problems reported.	No cure available to date. Symptomatic and supportive care management (nasogastric feeding, surgery).
			Myotonic dystrophy (DM)		
DM1	DMPK	19 (AD)	Any	Slowly progressing multisystemic disease leading to muscle wasting, cataracts, heart conduction defects, and myotonia. Endocrine changes reported. Respiratory problems infrequent.	No cure available to date. Symptomatic treatment and physical activity.
DM2	Zinc finger protein 9	3 (AD)	8–50		

MDC1A = congenital muscular dystrophy with merosin deficiency; MDC1C = congenital muscular dystrophy and abnormal glycosylation of dystroglycan; FCMD = Fukuyama congenital muscular dystrophy; WWS = Walker-Warburg syndrome; X-R = X-linked recessive; AD = autosomal dominant; AR = autosomal recessive.

Table 1. Summary of major muscular dystrophies.

With an estimated incidence of 1 in 3600 to 6000 boys, DMD is one of the most common and severe forms of muscular dystrophies with an early-onset and progressive muscle weakness leading to the loss of ambulation by the second decade. This X-linked recessive condition is caused by mutations in the dystrophin gene (*DMD*, locus Xp21.2), which is expressed in skeletal, smooth, and cardiac muscles, and hence the pathological involvement of different organs beyond skeletal muscle weakness, including respiratory and cardiac systems. The *DMD* gene comprises 79 exons and encodes a 14-kb mRNA transcript. This gives rise to a large protein product (427 kDa), a crucial cytoskeletal protein that mediates major structural and signalling functions within muscles. Four functional units make up the dystrophin protein. These include the actin-binding domain at the N-terminus, the rod domain consisting of 24 spectrin-like domains with four interspersing hinges, the cysteine-rich domain that mediates binding to β-dystroglycan, and the C-terminal domain mediating binding to syntrophin and dystrobrevin. These binding units confer a principal structural role for dystrophin facilitating its assembly with other proteins to form the dystrophin-associated glycoprotein complex (DAGC) (**Figure 1**) [1, 4, 5].

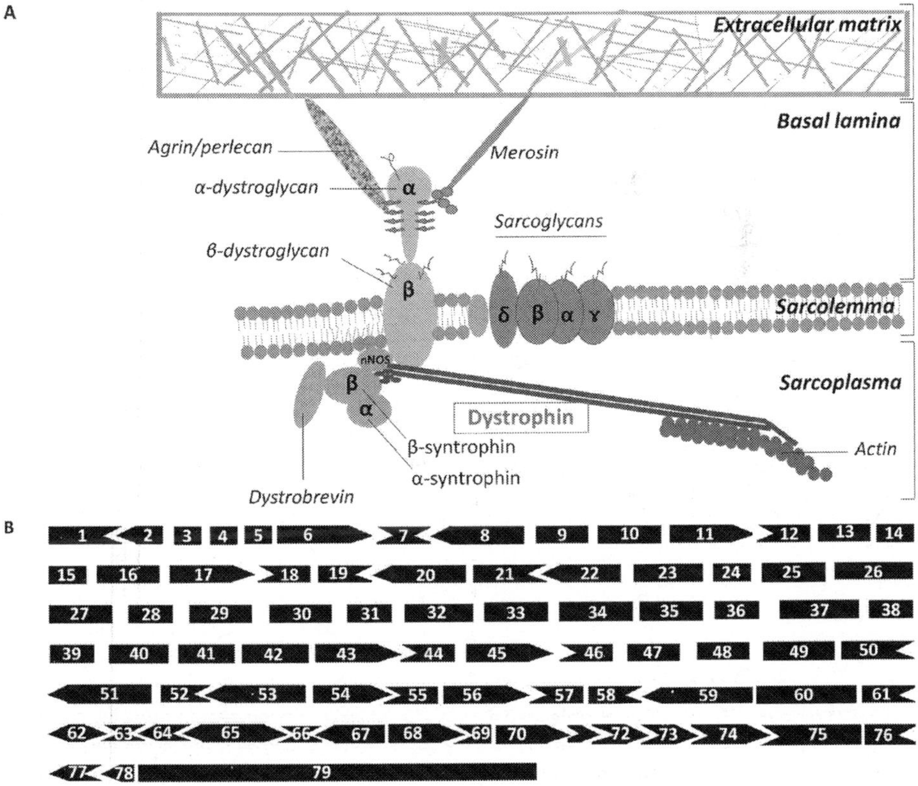

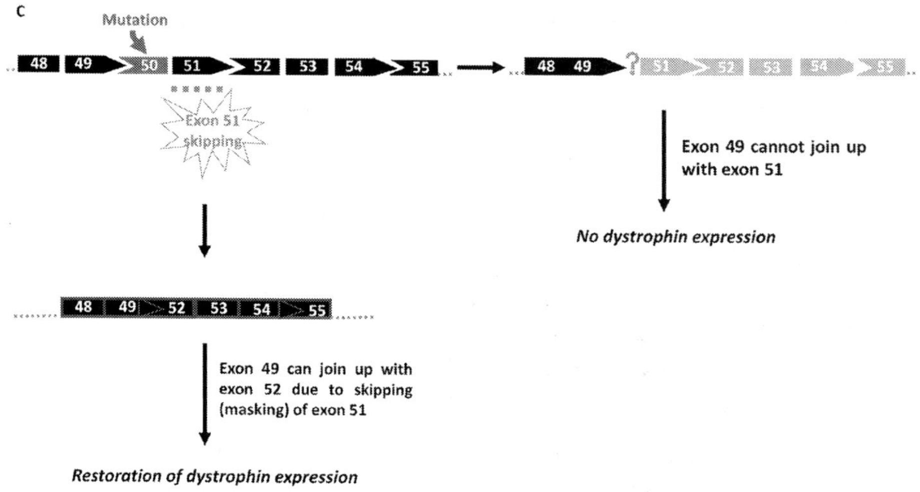

Figure 1. The role of dystrophin in muscle physiology and disease. (A) The Dystrophin-associated Glycoprotein Complex (DAGC) in skeletal muscle; (B) schematic illustration of exons that make up the *DMD* gene; (C) consequences of a frameshift mutation in the *DMD* gene illustrated by exon 50 deletion; and restoration of dystrophin expression with exon skipping therapeutics.

Deletions in one or more of the 79 exons that make up the *DMD* gene are common in DMD and account for approximately two thirds of the reported mutations. Other documented disruptions within the *DMD* gene include duplications (~10%), point mutations (~10%), and smaller rearrangements (15%) [6]. These lead to loss of dystrophin, which in turn destabilises the DAGC, resulting in the weakening of muscle fibre strength, increased susceptibility to stretch-induced damage, and raised intracellular calcium influx. These physiological disturbances account for the underlying histopathological features often observed in skeletal and cardiac muscles from affected patients, including muscle fibre necrosis, inflammation, and substitution with fibroadipose tissue [7, 8].

No curative treatment is currently available for most muscular dystrophies, including DMD. Current approaches involve relieving symptoms, delaying disease progress, and preventing complications [2]. Although these interventions proved beneficial in the short term, none of them can provide a long-term treatment and a permanent correction of the underlying pathological features. From a molecular point of view and based on advances in the identification of genes behind the observed phenotypes, most of these muscle pathologies represent good candidates for treatment with gene-based therapies.

3. Gene therapy for DMD

DMD has been the main focus and the proof-of-principle model for most gene therapy strategies targeting neuromuscular disorders, over the past years, with proof-of-concept validated in preclinical and clinical settings. The first clinical trial in the neuromuscular field, in fact, involved a gene replacement approach to deliver full-length dystrophin via a nonviral method of transfer [9]. The observed low expression levels, however, at local injections sites, highlighted the need for potent gene replacement transfer systems. Of these, adeno-associated viruses (AAVs) are seen as the best available option. The efficiency of functional dystrophin expression in patients with DMD, using AAV as a vector, was assessed in a phase I clinical trial [10]. The approach was shown to be safe with no concerning side effects, although overall efficiency was compromised by the development of dystrophin-specific T-cell-mediated immune responses. This finding was complemented by observations that preexisting T-cell-mediated immune responses to AAV were present, which, together with the dystrophin-specific T-cell responses, could have contributed to the low transgene expression levels detected following intramuscular injection. These early conclusions highlighted the need to circumvent immune destruction of therapeutic transgenes, by delivering unaffected homologs of dystrophin. Of these, a microutrophin-expressing recombinant AAV2/6 was shown to restore the dystrophin-glycoprotein complex and revert pathology in dystrophin (-/-)/utrophin (-/-) double-knockout mice model [11]. Delivering truncated versions of target proteins or their homologs, however, does not reconstitute a full gene replacement approach for DMD. Microdystrophin and minidystrophin transgenes often lack some crucial rod and hinge domains of full-length dystrophin, including neuronal nitric oxide synthase, syntrophin, and dystrobrevin, hence compromising maximal dystrophin functionality and membrane rigidity. This led to the engineering of triple AAV constructs using trans-splicing and hybrid methods, capable of delivering full-length dystrophin following coinfection of the tri-vectors in affected muscles *in vivo* [12, 13]. In trans-splicing, each vector acts as an independent construct holding sequential exonic sequences of human dystrophin's coding sequence. Coinjection of the vectors cause the constructs to cojoin via their inverted terminal repeats and deliver a full-length therapeutic transgene. These results circumvent the limited packaging capacity of AAVs and offer clinical hopes for boys with severe forms of DMD, for which the reversion of phenotype to a milder disease form [Becker muscular dystrophy (BMD)] using microdystrophin or minidystrophin-expressing AVV is not sufficient.

Although these gene replacement therapies are still in the early clinical phase of research and development, exon skipping-based therapeutics for DMD is progressing faster. Antisense oligonucleotides (AONs) have long been an effective alternative to dystrophin gene replacement therapy. These work by binding to the dystrophin transcript at sites that interfere with normal RNA processing, so that exons containing the mutations are bypassed, giving rise to in-frame transcripts capable of producing shorter yet functional protein products (**Figure 1C**). This was based on observations that BMD-like patients have in-frame transcripts with shorter dystrophin yet still maintain ambulation [14]. Hence, manipulating the splicing pattern of mutant dystrophin in DMD patients with an AON-based approach has the potential to alter disease phenotype from a clinically severe form of DMD to a milder BMD phenotype. For instance, the reading frame of dystrophin could be restored by blocking specific enhancers or splicing regulatory elements responsible for controlling the gene's exon recognitions. The approach was initially demonstrated *in vivo* in the *mdx* mouse model of DMD, bearing a single-point mutation in exon 23 that creates a stop codon with subsequent absence of dystrophin expression. Using a 2′-*O*-methyl (2′OMe) oligoribonucleotide complementary to the murine intron 22's 3′ splice site, it was possible to restore sarcolemmal expression of dystrophin in transfected myotubes *in vivo* [15]. The proof-of-concept has also been demonstrated in relevant human cell culture models derived from DMD patients bearing an exon 45 deletion, one of the most frequently deleted exons in DMD. The efficient restoration of dystrophin's coding frame was achieved by targeting splicing regulatory elements in exon 46 using a 2′OMe oligonucleotide [16]. Similarly, a successful correction was achieved through exon 51 skipping in a phase II clinical trial, where a successful restoration of both dystrophin and the DAGC was observed in patients with a deletion in exons 45 to 50 or exons 48 to 50 [8, 17], opening up great therapeutic hopes and paving the way towards late-stage clinical trials. A Biologics License Application (BLA) was submitted to the Food and Drug Administration (FDA) for approval of the antisense agent PRO051, 2′OMe (drisapersen) targeting exon 51 for DMD. However, this application was recently rejected, as a phase III study (NCT01803412) of long-term intake failed to meet its primary efficacy endpoint and also showed evidence of significant toxicity to a number of organs. Another similar antisense agent, eteplirsen, is currently under the FDA review for approval.

Although AON-based therapy for DMD has long been the prime attention of scientists and clinicians, this approach does not provide a long-term cure for DMD. Regular administrations of high doses are required to achieve a constant skipping and redirection of gene expression.

4. Gene editing strategies

With advances in human genome and increasing need to provide a simplified long-term curative approach for genetic muscle diseases, new sophisticated technologies based on gene editing have emerged. These aim to permanently correct disease phenotypes in affected individuals using site-directed endonucleases such as zinc finger nucleases (ZFNs), transcription activator-like effector nucleases (TALENs), and clustered regularly interspaced short palindromic repeats (CRISPR). These corrections occur through the generation of double-stranded DNA breaks that will eventually induce intrinsic cellular DNA repair mechanisms mostly via nonhomologous end joining (NHEJ) [18, 19].

Zinc finger nucleases (ZFNs)

ZFNs are currently the highly advanced gene editing system. Initially, these were rationally designed taking advantage of key biological properties of zinc finger transcription factors such as their DNA sequence recognition function. Each zinc finger component is composed of ~30 amino acids and is capable of recognising 3 bp of DNA. The overall structure is arranged in tandem repeats of zinc finger motifs, hence allowing the recognition of longer sequences of DNA. The nuclease activity of ZFNs is conferred through fusion to *Fok*I endonuclease's catalytic domain. This design ensures that enzymatic and subsequent DNA cleavage activities are only targeted to sites recognised by the binding domains of ZFNs. The latter consist of Cys2His2 zinc finger structures, in which a single zinc atom is surrounded by 30 amino acids and is capable of recognising 3 bp of DNA. More often, however, three to six zinc finger units

http://dx.doi.org/10.5772/62282

are assembled, which allows the recognition of 9- to 18-bp DNA sequences. This is usually regarded as acceptable for a single locus targeting in a human genome. From a structure-activity relationship, however, full functionality is not achieved unless *Fok*I is presented in a dimerised form so as two DNA cleavage domains are dimerised around the target DNA sequence [19, 20]. In an attempt to maximise the biosafety profile of ZFNs, several methods have been employed to increase specificity. These include limiting the spacer length between the recognition sites of chimeric ZFN subunits [20]. In fact, ZFNs with shorter interdomain linkers connecting the Cys_2-His_2 zinc finger and the nuclease domains were shown to have a restricted activity with a 6-amino acid linker exerting the most selective activity at a target DNA site with a 6-bp spacer [21]. Another approach has been to generate obligate heterodimer nuclease domains with decreased off-target effects [22–24]. This is associated with a relatively weaker interaction between the heterodimer cleavage domains, which would necessitate stronger interactions between each monomer and the target site to achieve a site-specific cleavage while minimising cleavage at weakly bound nontarget sites. This low-affinity high-avidity approach has been proposed as a plausible mechanistic approach behind the site-specific cleavage activity of the newly engineered obligate heterodimer *Fok*I domains. An alternative to the obligate heterodimer method has been the expression of autonomous ZFN pairs, the combined expression of which was as effective as obligate heterodimer ZFN domains at inducing targeted chromosomal deletion in mammalian cells, with reduced toxic effects that are often thought to be linked with unwanted individual ZFN subunits' cross-reaction [25]. The enhancement of *Fok*I's enzymatic activity has also been reported, whereby an *in vivo* evolution-based method was employed to further increase the cleavage activity of *Fok*I. The incorporation of the enhanced domain in heterodimer ZFN structures resulted in a potent product with improved overall cleavage profile [26].

The successful correction of different genetic mutations associated with various diseases, including sickle cell anaemia [27], haemophilia [28], α1-antitrypsin deficiency [29], X-linked severe combined immunodeficiency [30], and, more recently, HIV [31, 32], has led to its application in DMD, in which precise gene editing can be achieved by deleting targeted exons from the dystrophin gene. For instance, using extended Modular Assembly (eMA)/Context-Dependent Assembly (CoDA) methods, it was possible to generate several exon 51 targeted ZFNs. Two of which demonstrated a remarkable activity with mild off-target mutagenic effects in myoblast cells from DMD patients, harbouring a deletion of exons 48 to 50. When implanted into the hind limb of immunodeficient mice, the corrected myoblasts were capable of maintaining dystrophin expression *in vivo*, with a correct sarcolemmal localisation [33]. This proof-of-concept study has shown that the ZFN-based approach could potentially be adopted under a cell-based therapy approach for DMD, hence holding a promising faster translation into the clinic, considering that the approach is already in clinical trials for *ex vivo* cell modifications in HIV [32].

Transcription activator-like effector nucleases (TALENs)

TALENs were originally isolated from *Xanthomonas* bacteria, plant pathogens capable of employing up to 40 effector proteins to circumvent eukaryotic cell defences during a host infection [34]. These are composed of repeated motifs of 33 to 35 amino acid residues, identical with the exception of the 12th and 13th residues, which are often known as the repeat-variable di-residues. The latter play a key role in TALEN's DNA-binding specificity, whereby a different pair of amino acids would exhibit a specific binding to a corresponding nucleotide in the target sequence. For instance, the asparagine (Asn; N)-isoleucine (Ile; I) NI, the histidine (His; H)-aspartate (Asp; D) HD, the Asn-Asn NN or the Asn-lysine (Lys; K) NK, and the Asn-glycine (Gly; G) NG pairs preferentially bind to adenine, cytosine, guanine, and thymine, respectively. These constitute TALENs' DNA-specific binding domains, which, like ZFNs, are conjugated to nonspecific *Fok*I cleavage domains, hence directing them to the target site for gene editing and subsequent correction of the final protein product. A recognition sequence of 14 to 20 bp, together with an appropriately spaced *Fok*I subunits separated by 12 to 19 bp, will ensure good genomic target recognition with maximal cleavage activity owing to efficient *Fok*I dimerisation [35, 36].

Although not yet being as clinically advanced as ZFNs, the lower cost and ease of production of TALENs have generated an increasing interest in the technology. Recent work in DMD has shown the feasibility of the approach using optimised exon 51 targeted TALEN-encoding plasmids transfected into myoblast cells isolated from two different patients with deletions in exons 48 to 50. A satisfactory gene correction was observed with up to 12.7% and 6.8% of alleles confirmed to have indels in the two treated patient myoblast cell lines, respectively. This correction was further correlated with a good restoration of dystrophin expression at the expected predicted size of ~412 kDa compared to its expression in the wild-type isolated myoblasts [37]. Whole exome sequencing of the successfully corrected TALEN-treated cells revealed no insertions or deletions, except at the exon 51 target locus. Further analysis using the TALE-NT 2.0 Paired Target Site Prediction web server confirmed the nature of the observed single-nucleotide variants occurring. These were in fact related to expected genomic mutations that normally occur during cell clonal expansion, as none of these showed any similarity to the employed TALEN target site with spacers of 1 to 30 bases [37].

Besides the dystrophin gene as a main target in DMD, other genes have received comparable interest as targets for TALEN-based gene editing. Of these, myostatin (MSTN) has been the centre of attention as a member of the transforming growth factor-β family. This is mostly linked to its recognised role in muscle physiology, acting as a negative regulator of skeletal muscle mass. In fact, studies have shown that mutations in the *MSTN* gene cause an increase in skeletal muscle fibre numbers and sizes, which in turn lead to muscle hypertrophy without any alarming consequences [38–40]. These observations were recently translated into therapeutic interventions, whereby inhibiting the MSTN signalling pathway using pharmacological agents has shown real benefits in DMD but also in other muscle wasting pathologies such as sarcopenia and cancer cachexia [41–45]. A phase I study is currently under way to assess the safety of an anti-MSTN monoclonal antibody in advanced cancer patients with cachexia (NCT01524224). The same approach using a different pharmacological agent, BMS-986089 (Bristol-Myers Squibb), will soon be tested in a first clinical trial involving patients with DMD (NCT02515669). From a molecular therapeutic point view, *MSTN* gene editing provides a long-term control to switch off MSTN signalling in the wasted affected muscles. The approach was successfully demonstrated recently using a pair of TALEN-expressing plasmids targeting the human exon 2 locus, a highly conserved region within the coding sequence of the *MSTN* gene. Consistent with previous ZFN-related toxicity studies, the reported MSTN-TALEN targeted system was engineered using obligate heterodimers of the *Fok*I domain to minimise off-target effects often seen with homodimer variants. Initial experiments in the HEK293 cell line revealed that the mutation induced by the TALEN approach was persistent with indels still detected 1 month after transfection using the T7E1 assay, a commonly used enzyme mismatch cleavage method for detecting mutations. A similar finding was reported in four different cell lines from different species including human, bovine, and murine cells. Further investigation in primary myoblast cultures derived from a dysferlin-deficient mouse model as well as from patients with dysferlinopathy confirmed an efficiency of 10.3% to 24.6% of gene editing after treatment [46]. Although most ZFN- and TALEN-based gene editing systems for muscle diseases to date have been engineered based on NHEJ mechanisms, proof-of-concept data are now emerging on TALEN-mediated homology-directed DNA repair (HDR). Targeted integration of dysferlin, a mutation of which is associated with limb-girdle muscular dystrophy (LGMD) type 2B and Miyoshi myopathy, has been shown following cotransfection of HEK293 with MSTN targeted TALEN-expressing plasmid and a donor plasmid expressing dysferlin tagged to an enhanced cyan fluorescent protein under a cytomegalovirus (CMV) promoter [46]. Although this approach is still at its early research and development stage, the results obtained *in vitro* remain promising and could offer the basis for future TALEN-corrected myoblast transplantation therapy for a number of severe muscular dystrophies.

Clustered regularly interspaced short palindromic repeats (CRISPR)

Recent achievement in the field of gene editing has seen the emergence of CRISPR as a novel correction tool in biomedicine. CRISPRs were first described in prokaryotes as part of an RNA-guided adaptive immune system to protect against foreign intrusions such as viruses and plasmids. In response to these, bacteria and archaea are programmed to incorporate short sequences of foreign DNA at one end of a repeat element, which is often referred to as the CRISPR. The integrated inserts between the repeat elements of prokaryotic CRISPR-associated (Cas) loci hence confer a permanent future mechanism of defence against the past invaders [47]. Three types of CRISPR/Cas systems (I–III) have been described, with differing structural, functional, and mechanistic characteristics. Each system, however, shares a similar sequence arrangement of short repeated units of 30 to 40 nucleotides, separated by a unique "target" specific nonrepeated sequence (a spacer) of equal length [48, 49]. A leader sequence incorporating a promoter is often present, which initiates a unidirectional transcription of CRISPR sequences. Whereas types I and III systems are present in both bacteria and archaea, type II systems have been reported in bacteria only [50]. Type II systems that use a unique Cas protein referred to as Cas9 are of particular interest in medical application. The RNA-guided, nuclease-mediated genome editing of a type II CRISPR system is mediated by Cas9, a nuclease that is directed to the target DNA site by a single-guide RNA recognising a specific locus next to the protospacer adjacent motif. This subsequently creates a double-stranded break, which could be repaired by NHEJ or by a homology-directed mechanism of repair provided that an exogenous plasmid donor is codelivered [51].

In muscular dystrophy, the CRISPR/Cas9-based technology could theoretically be employed to correct germ-line DNA prenatally in one-cell zygotes or as a postnatal treatment. The first approach was recently tested *in vivo* in *mdx* zygotes. By injecting a single-guide RNA-guided CRISPR/Cas9 together with an appropriate HDR template, Long et al. demonstrated for the first time the feasibility of the approach to correct a nonsense mutation in exon 23 of the *DMD* gene [52]. Although the CRISPR/Cas9-treated *mdx* zygotes produced mice with varying degrees of gene correction, those showing 41% of gene correction by HDR and 83% correction by NHEJ repair mechanisms demonstrated a complete disease-free phenotype with normalisation of dystrophin expression at the histological level [52]. Despite showing some promising results in DMD, a successful germ-line DNA gene editing strategy would (if ever ethically permissible) require a good knowledge of the nature of mutation affecting the maternal disease carrier to allow a site-specific targeted correction. This prior knowledge, however, could be problematic in X-linked DMD in which it is estimated that one third of all mutations in the *DMD* gene arise spontaneously [53] and hence would not qualify for a Cas9 gene editing-based treatment. This would restrict the approach to the correction of well-characterised known point mutations, which only make 15% of DMD mutations. One alternative approach to consider would be a postnatal *DMD* gene correction. This was recently demonstrated using multiplexed single-guided RNAs to direct the Cas9 nuclease to mutations at exons 45 to 55 of the dystrophin gene [54], an approach that has the potential to correct more than 60% of mutations in DMD patients owing to the large hotspot mutation deletion achieved. Despite a good restoration of dystrophin expression both *in vitro* and *in vivo* following transplantation of the CRISPR/Cas9-treated DMD patient myoblasts into immunodeficient mice, the overall deletion efficiency, however, was less than that obtained following exon 51 deletion [54]. This suggests that the size of the targeted sequence is crucial in dictating Cas9-associated gene editing capacity to mediate an efficient repair by NHEJ or HDR, where a size-dependent decrease in nuclease-mediated gene deletion was previously noted [55]. Although proof-of-concept has eloquently been shown in research settings, CRISPR/Cas9-based gene editing in DMD would only be envisioned in the clinic should appropriate gene delivery methods are employed. Gene transfer systems capable of directing the elements of the CRISPR/Cas9 cassette to the diseased muscles *in vivo* are needed to ensure a satisfactory *DMD* gene editing and a full therapeutic outcome. For instance, AAV has previously been shown to be a safe and an effective gene transfer system in some clinical trials for gene replacement therapy [56]. In this regard, the AAV8 and AAV9 serotypes would be useful delivery tools for CRISPR/Cas9-mediated gene editing for DMD, owing to their reported efficient gene transfer to skeletal muscles and the heart following systemic administration [57, 58], which therefore could potentially result in a robust gene correction at the key target tissues harbouring the mutation in DMD patients. Recent studies have demonstrated the efficiency of AAV as a good platform tool for CRISPR/Cas9-mediated gene editing transfer to dystrophic muscles in DMD disease models *in vivo*. The approach was recently tested following administration by three different systemic routes, including intraperitoneal, intramuscular, and retro-orbital injections in postnatal mdx mice, and proved to be successful with an increasing dystrophin expression reported from weeks 3 to 12 after virus injection. This correlated with an overall improvement in skeletal muscle function [59]. Mutated exon 23 targeted deletion was also demonstrated to restore disease phenotype in both neonatal and adult DMD models *in vivo* by systemic and local delivery of a CRISPR/Cas9-expressing AAV [60]. A similar viral-mediated transfer of a CRISPR/Cas9 system coupled with paired guide RNAs flanking the mutated exon 23 in DMD proved that AAV is a good gene editing transfer system for local and systemic restoration of dystrophin expression in muscle cells but also myogenic stem cells in diseased muscle *in vivo* [61].

5. Application in trinucleotide repeat expansion muscular dystrophies

Proof-of-concept studies in DMD have opened new horizons in the application of gene editing in other muscular pathologies for which the correction of toxic genes is believed to address the underlying dynamic mutations often associated with triplet repeat expansion and some locus contractions disorders.

Oculopharyngeal muscular dystrophy

OPMD is an inherited autosomal dominant, slow-progressing, late-onset degenerative muscle disorder characterised by progressive eyelid drooping (ptosis), swallowing difficulties (dysphagia), and proximal limb weakness. Whereas the incidence in Europe is 1/100,000, the disease has been largely reported in the Bukhara Jew population in Israel (1/700) and the French Canadian population in Quebec (1/000) [62].

The underlying genetic defect behind OPMD is an abnormal expansion of a (GCG)n trinucleotide repeat in exon 1 of the PABPN1 gene, which leads to an expanded polyalanine tract at the N-terminal of the PABPN1 protein (12–17 repeats are often detected in mutant PABPN1 compared with only 10 repeats in the wild-type protein) [62].

From a molecular therapeutic point of view, OPMD is a good candidate for gene editing-based treatment. Previous studies have shown the importance of abrogating intranuclear inclusions (INIs) of mutant PABPN1, the main pathological hallmark of the disease, with neutralising agents capable of binding aggregated mutant proteins. Of these, intrabodies were shown to be effective in reducing mutant PABPN1-associated INIs and restoring a normal pattern of gene expression in a *Drosophila* OPMD model [63]. Although these show antiaggregate properties by tackling the disease phenotype at the protein level, they do not address the underlying molecular pathology behind it, hence necessitating a repeated administration approach to keep the disease under control. In this regard, gene editing could offer a long-term permanent therapeutic advantage. Allele-specific correction of expPABPN1 can be achieved by CRISPR/Cas9-mediated editing targeting the expanded GCG moiety. Unlike recently tested short hairpin RNA (shRNA)-based knockdown of PABPN1 [64], this gene level of correction could result in a permanent production of functional PABPN1 protein that is capable of correcting the OPMD phenotype at both histological (reduction of pathological aggregates) and molecular (abrogation of mutant PABPN1) levels. The need for an adjunct gene replacement therapy as is often the case in knockdown approaches [62] could no longer be necessary. However, thorough characterisation and mechanistic studies would need to be conducted both *in vitro* and *in vivo* to confirm CRISPR/Cas9 editing specificity towards the expanded mutant

PABPN1 with no obvious off-target effects on the wild-type PABPN1 gene or indeed other genes within the treated tissues.

Myotonic dystrophies

DM is an autosomal dominant, slow-progressing inherited multisystem genetic disorder affecting the muscles (causing wasting), the eyes (leading to cataracts), and the heart (causing conduction defects). The disease is also characterised by metabolic disturbances (endocrine changes) and prolonged contraction of skeletal muscles (myotonia).

Two types of DM have been defined to date. These include DM type 1 (DM1) and DM type 2 (DM2), with DM1 being the most severe and most common form of DM. Overall incidence has been estimated at 3 to 15 per 100,000 in Europe with a higher prevalence in Iceland (1:10,000) and a reported incidence of as high as 1:500 in Quebec [65]. The disease is caused by a CTG trinucleotide expansion in the 3′-untranslated region of the DMPK gene. CTG repeats exceeding 37 are considered abnormal with larger repeats (below 400 repeats) leading to a more severe disease [66, 67]. Like most debilitating muscular pathologies, no curative treatment currently exists for DM. The use of pharmacological agents [68, 69] and antisense RNA-based therapies [70–72] have shown improvements in disease phenotype, but overall benefits on the long-term remain, in some cases, limited and full therapeutic efficacy is yet to be demonstrated in the clinic. Although still in its early infancy, gene editing for DM treatment is an appealing alternative to most pharmacological and gene therapy approaches reported to date. Correction of mutant DMPK at the gene level offers the possibility for a permanent modulation of pathological phenotypes and long-term disease control. A recent study has shown the feasibility of editing intron 9 in the DMPK gene in DM1 neural stem cells derived from human DM1 induced pluripotent stem cells (iPSCs) using a TALEN-mediated homologous recombination-expressing cassette integrated upstream of the CTG repeats [73]. A significant reduction in nuclear RNA foci, together with restoration of normal microtubule-associated protein τ (MAPT) and muscleblind-like (MBNL) splicing patterns, were observed [73]. Although transition into clinical use may be long and difficult, this proof-of-concept *ex vivo* study offers a rationale for the genetic correction of DM1-dervied stem cells as a potential autologous cell therapy for DM1 patients in the future.

6. Regulatory challenges and pathways into the clinic

In October 29, 2015, the United Kingdom became the first country in the world to legally approve one type of a human germ-line gene modification based on mitochondrial replacement. This approach is thought to save at least 10 children each year from mitochondrial diseases by preventing maternal transfer of mutations in mitochondrial DNA to offspring (Department of Health, 2014). Human germ-line genetic engineering, however, is currently not permitted in the United Kingdom and in other European countries. Although a somatic gene editing-based approach would be less questionable from an ethical point of view, a number of safety concerns would need to be addressed from a regulatory perspective. In

http://dx.doi.org/10.5772/62282

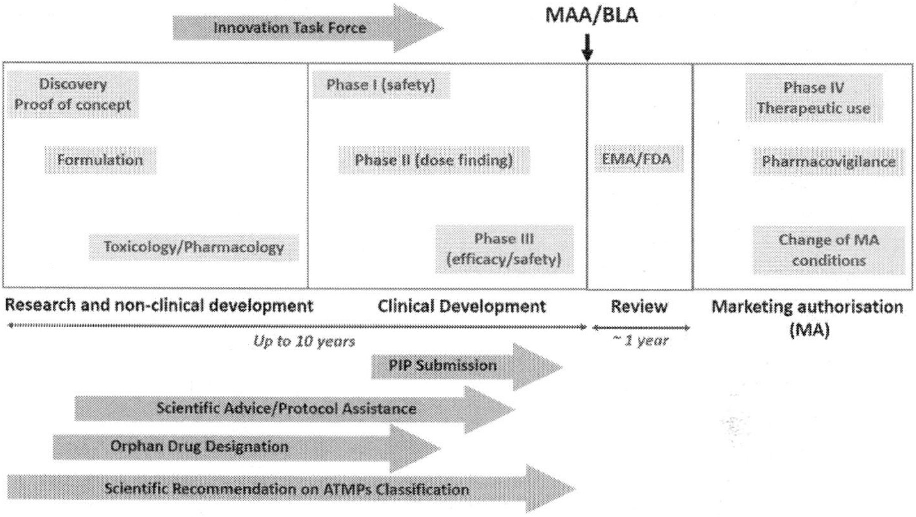

Figure 2. Regulatory pathway for advanced therapy medicinal products. MAA = Marketing Authorisation Application; BLA = Biologics License Application; EMA = European Medicines Agency; FDA = U.S. Food and Drug Administration; PIP = Paediatric Investigation Plan; ATMPs = Advanced Therapy Medicinal Products.

accordance with Article 2(1)(a) of Regulation (EC) No. 1394/2007 and as per the European Medicines Agency (EMA) classification, gene editing-based products, including cells modified *ex vivo*, will be regulated as advanced therapy medicinal products (ATMPs) for which a central authorisation procedure governed by the EMA would apply (**Figure 2**). This would eventually lead to a single marketing authorisation that is valid across the entire EU and the European Economic Area (EEA) countries. A number of guidelines on the requirements for product quality and preclinical and clinical studies have been issued by the EMA over the past years to facilitate the transition of promising experimental advanced therapies into the clinic.

Safety considerations for gene editing

The risk of inducing modifications at off-target genes, leading to unwanted side effects, is a major safety concern. The degree of off-target events and their clinical implications are crucial questions that need to be carefully addressed as part of a regulatory new investigational drug development plan. The primary concern is that modifications could be carcinogenic. It must also be remembered that (unless the therapy is based on a cell type that is selected and clonally expanded after modification) the off-target effects will be heterogeneous from cell to cell, so the analysis must encompass a suitably large (and potentially diverse) population of cells exposed to the modifying agent to evaluate if any cells may be adversely transformed even at low frequency. There may also be risks associated with changes that have cytotoxic effects, and if such effects are frequent in the same cells that undergo correct gene editing, the consequences could include loss of efficacy.

In this regard, rigorous quality assurance tests would need to be conducted to identify any insertions or deletions that could result from a gene editing NHEJ-based treatment. These include gene sequencing, mismatch cleavage assays based on CelI or T7 endonuclease I enzymes, and the tracking of indels by decomposition (TIDE) method. These tests could be used as quality assurance tools to determine the number and location of indels or substitution events occurring after gene editing treatment, more often based on *in silico* predictions. However, as previously noted with ZFNs, some mutations could occur near cryptic off-target sites that are not predictable *in silico* [74]. Care should also be taken when validating the assays employed for product characterisation purposes. This follows from previous observations whereby initial analysis by whole genome sequencing (WGS) in gene-corrected human iPSCs (hiPSCs) revealed a large number of indels, of which some were confirmed to be false positive [75]. This highlights some limitations of WGS and calls for a multitesting approach employing different analytical methodologies.

Although current assays would be valuable in identifying the location of suspected off-target mutations that could arise from NHEJ or HDR-based gene editing treatments, they do not, however, provide precise information on the ability of these mutations to cause carcinogenicity in the long term. Hence, additional functional studies are required to assess the significance of these off-target events at the molecular and cellular levels. For instance, an unwanted insertion in the middle of an essential gene sequence could have dramatic consequences. One should have enough knowledge on the biological and physiological functions of the affected gene to draw an informed decision on the overall safety of the proposed approach. Similarly, an off-target insertion into an enhancer or a repressor region would disturb genetic homeostasis resulting in an unwanted down-regulation or up-regulation of genes that come under the affected promoter. This could subsequently interrupt normal cellular activity with potential alteration of phenotype at histological and physiological levels. Although not all mutations could affect cellular proliferation, an unwanted "indel" mutation occurring within genes known to regulate cell replication or involved in programmed cell death raise an alarming concern from a safety point of view. This is due to potential risks arising from a compromised cellular viability leading to severe toxicity in healthy tissues or an uncontrolled replication that could eventually lead to tumour formation. These potentially serious consequences warrant careful considerations during early product development stages; hence, basic biological studies to further characterise the sites at which these mutations have occurred should be conducted as ad hoc validation studies to rule out any unwanted insertional mutagenesis and/or tumourigenic consequences following a gene editing-based treatment.

Although clinical safety experience with biological therapies based on gene editing is lacking, current product development and regulatory strategies should also draw from past and available human clinical trial data on retroviral and lentiviral vectors' integration sites. A key safety parameter would be to satisfy the regulatory bodies that those detected off-target effects have been thoroughly characterised and occur at "low risk" sites unlikely to cause insertional oncogenesis following integration of foreign DNA. In fact, not all events at off-target genomic sites would result in serious side effects. For instance, those occurring at "extragenic" sites distant from essential gene regulatory sequences are less likely to cause harm than those affecting "intragenic" sites. Modifications to sites that do not fall within a gene transcription unit as well as those situated more than 50 kb away from the 5′ end of any gene and more than 300 kb from genes linked with cancer or microRNA sites are generally considered safer. Genomic sites not within ultraconserved regions and outside long noncoding RNAs would also represent low-risk sites in the event of off-target DNA integration [19, 76].

Furthermore, validation studies should not only focus on the affected off-target sites but also neighbouring sites by measuring the effect of these mutations on neighbouring gene expression. These studies should ideally be validated *in vivo*. The choice of appropriate animal models, however, remains a challenge and care should be taken when interpreting data [19, 76, 77].

Validation and choice of functional studies for assessment of toxicity

Although basic biological studies are valuable in assessing the extent of genome modification as a whole and increasing our understanding of different alterations taking place including those that are unlikely to result in clinical toxicity, they do not fill all the regulatory requirements gaps. Functional toxicity studies would, therefore, need to be conducted in parallel early during the preclinical stage of product development. Two paramount questions would need to be carefully addressed to satisfy regulators' concerns on serious toxicity issues that are specific to genome editing-based medicines. These include cytotoxicity and genotoxicity. Tumourigenicity is the most important safety consequence to consider, but germ-line effects are also not expected to be permissible. Cytotoxicity may be caused by (i) the vector itself (including expression of viral vector antigens and potential persistence of the vector in cells), (ii) gene editing machinery (and any persistence thereof, especially DNase), (iii) off-target gene editing, or (iv) on target gene editing (possible if the editing is either not accurately restoring the wild-type or is generating new antigenicity and/or genetic instability).

Eloquent studies and approaches would have to be designed to address these issues. At preclinical stage, this could include viability assays based on GFP-positive (GFP^+) cells, whereby the treated cells are cotransfected with the tested nuclease-expressing construct and a GFP-expressing plasmid. This would allow investigators to track and quantify any observed decline in the GFP^+ cell population as a result of a nuclease-related toxicity. The approach would also be valuable in dose escalation studies when deciding on optimal dosage administration for subsequent first-in-human clinical studies. Monitoring of clonal changes has also been used as an informative way to assess genotoxicity *in vitro* in cells pretagged with unique short sequence identifiers to allow tracking of changes in starting clone dynamics over a period of time [78]. Similarly, other approaches have relied on fluorescently tagged cell cycle indicators in cell lines such as HeLa FUCCI cells as a complementary *in vitro* validation method to assess the genotoxicity effects of genome editing at the cell cycle level [79].

However, it is important to bear in mind that measuring cytotoxicity and genotoxicity using approaches based on reporter genes and tagging systems is restricted to preclinical assessments and cannot be employed at later stages of product development when assessing potential toxicity in human clinical trials. Hence, the need for clinical toxicity assays that are fit for purpose should not be neglected. Furthermore, correlation of these assays with clinical outcomes is yet to be demonstrated. For this reason, all functional toxicity studies conducted *in vitro* should use cellular models and gene transfer methods that are similar to those intended to be applied in clinical settings to fulfil some of the safety regulatory requirements.

Vector considerations for in vivo and ex vivo gene editing

Potential cytotoxic effects do arise not only from on-target and off-target effects but also from the employed gene transfer system itself. In this regard, the type of vector (viral or nonviral) and the gene (nuclease) transfer approach (direct *in vivo* delivery versus *ex vivo*) adopted will have to be implemented earlier during product development and considered as part of the constructed regulatory strategy for the nuclease-based medicine under consideration.

For severe neuromuscular disorders in which there is a widespread distribution of affected muscles, AAV vectors and in particular serotypes 8 and 9 are seen as ideal systems for gene editing-based nuclease transfer to diseased muscle tissues owing to their relatively high tropism for skeletal muscle cells and the heart [57, 58] as well as their documented safety profile in the clinic following recent approval of Glybera® in Europe [80]. From a product development point of view, this can be achieved through an HDR-based gene editing construct packaged within a recombinant AAV. Although technically this is achievable with ZFNs due to a relatively small monomer insert size of ~ 1.2 kb, this might not be the case for Cas9 nucleases. With an insert size of ~4.1 kb, this could be an issue when it comes to packaging a bicistronic construct harbouring a donor DNA template [81], considering the limited packaging capacity of rAAVs. Taking into account the EMA's guidelines on quality, nonclinical and clinical issues related to the development of recombinant adeno-associated viral vectors (EMEA/CHMP/GTWP/587488/2007 Rev. 1) [82], caution should be taken during the manufacturing stage of the vector to avoid the possibility of producing AAV particles whose packaged DNA is greater than that of wild type virus. Technically, this could be overcome by splitting the Cas9 gene between two vectors. However, this markedly increases the developmental and regulatory complexity, with each product warranting a full characterisation and a thorough assessment of approaches employed during manufacturing, quality control, and preclinical evaluation (including choice of animal models for testing, vector persistence and tropism, reactivation of virus infection, and germ-line transmission) in addition to clinical studies. The latter would have to be based on a data-driven dose selection for each vector used and show a comprehensive picture of virus biodistribution and shedding, immunogenicity profile, and germ-line transmission considering the permanent gene modification achieved with gene editing-based nucleases and the risks associated if the virus expressing these genes accidentally infect germ-line cells. A long-term follow-up is therefore highly recommended and should not be neglected as part of a complete regulatory plan for a gene editing-expressing AAV vector.

Similarly, the use of genetically modified and nonmodified cells including myoblast cells and myogenic stem cells for the treatment of monogenic inherited neuromuscular diseases has been well validated in preclinical and clinical studies [83–86] and it is anticipated that these therapies would become normal treatment modalities in the clinic in the future.

Although gene edited-based cell therapies for HIV and leukaemia are seeing a rapid positive progress in clinical trials and are rapidly extending to other debilitating diseases, a smooth and safe progress of these life-changing advanced therapies in the market would require fulfilments of a number of criteria from a regulatory point of view. The ultimate goal is to ensure that these therapies would reach the severely affected individuals at minimal health and safety risks with regards to the treated patients themselves as well as third parties and the environment in large. According to the EMA's guidelines, a multifactorial risk assessment approach needs to be considered taking into account the origin of cells involved, the type of vector employed during the genetic modification procedure, the manufacturing process, the noncellular components used as part of the formulated product, and the intended specific therapeutic use of the final product.

Lentiviral vectors have long been the vector of choice for the *ex vivo* engineering of stem cells and muscle progenitor cells for the treatment of muscular dystrophies [87–89]. This adds an additional layer of regulatory requirements from quality, nonclinical and clinical perspectives. The manufacture of viral vectors to the GMP quality standards required for an approvable medicinal product is expensive and often inefficient, but developers are helped by the guidance issued in these areas by the EMA (for example, "Guideline on the quality, nonclinical and clinical aspects of gene therapy medicinal products (EMA/CAT/80183/2014)") [90], as well as the monograph of the European Pharmacopoeia (Ph. Eur. 5.14). A thorough product characterisation is required as part of this process that would include different but complementary methods based on molecular, biological, and immunological assays with the aim of assuring the identity, purity, and potency of the produced genetically modified cells as a final product. Major considerations for viral vectors used *in vivo* are the biodistribution, potential for generation and/or shedding of infective virus, potential for germ-line modification, and the impact of interaction with the recipient's immune status. Clinical use of viral vectors and genetically modified cells should also comply with applicable regulations for genetically modified organisms.

Complexities arise from the genetic manipulation (gene correction) as well as the differentiation status and capacity of the modified cells, which could result in a mosaic cell population. Furthermore, intrinsic variations between cells as a result of donor differences give rise to massive batch variations in the final product. From a regulatory perspective and in line with the EMA's guidelines, nonclinical and clinical studies need to be conducted with cell medicinal products that are well characterised. The manufacturing process needs to be robust and quality control focused capable of maintaining consistency and reproducibility of the final cell-based product. For this, all starting materials need to be well defined and carefully documented. For treatment of muscle diseases, it is often reasonable to administer cell-based products that are in a differentiated state, which may pose less tumourigenic risks. However, one cannot exclude the existence of a subpopulation of cells in an undifferentiated proliferative state. For iPSCs, for example, it is empirical to conduct additional testing of cell transformation and tumour formation during the early manufacturing stage of the product as a precaution measure. This would often need to be combined with the selection of appropriate markers during critical manufacturing steps for assuring a defined stage of differentiation that is intended for therapeutic use.

For any cell-based product that has undergone a substantial *ex vivo* manipulation, a robust process validation process is therefore paramount and should include a combination of genetic stability, tumourigenicity, and phenotypic profile assessments of both wanted and unwanted cell populations at all critical stages of manufacturing to ensure safety requirements are met. The mosaic nature of cell-based medicinal products often complicate their identity and purity-related characterisation. Most often, it is reasonable to accept that purity does not always equate homogeneity. Truly selective markers that could accurately map and distinguish different cell types and differentiation stages are yet to be identified, which often render product characterisation a challenging task for most developers. Nevertheless, one should not underestimate the importance of a thorough demonstration of product consistency as a minimum requirement for characterisation purposes.

The ability to track any cell-based therapy following administration in patients is crucial for clinical monitoring purposes. However, limitations in current medical methodologies do not allow a full biodistribution profile to be drawn from human studies. Hence, the importance of thoroughly designed biodistribution studies in nonclinical models should take into account the multistep biodistribution characteristics of cell products, including migration, niche, engraftment, differentiation, and persistence, together with reliable *in vivo* tracking methods such as the use of marker genes or labelled cells. Current European regulatory requirements do not give exemptions when the risk profile of the cell-based product under investigation is subject to a safety concern or when its route of administration (such as intravenous delivery) warrants a special attention. For this, noninvasive methods based on clinically accepted tracers should be considered and their use should be justified when conducting biodistribution studies in human trials.

7. Conclusion

For diseases caused by genetic defects, means to correct the defect are attractive routes to cure the disease at its source. Medical scientists working in the field of muscular dystrophies are therefore excited by the opportunities that recent developments in gene therapeutics bring to the field. Technologies do now exist for manipulating the human genome in *ex vivo* cultured cells with considerable specificity. However, there are further technical challenges to solve before such technologies translate into viable medical treatments for affected individuals. The medical need is evident from the severity and incidence of these conditions, which are frequently severely debilitating. It should also be noted that muscular dystrophies are not a single disease but cover a range of conditions caused by different mutations, and it is to be expected that, for specific gene-based therapeutics, a different medicinal product will generally be required according to the different mutation that has caused the disease.

The increasing number of investigational medicinal products entering clinical trials involving patients with DMD and BMD has prompted the EMA to publish its guidelines on the clinical

http://dx.doi.org/10.5772/62282

investigation of medicinal products for the treatment of DMD and BMD (EMA/CHMP/236981/2011, Corr. 1) [91], expected to come into effect in July 2016. This gives a general guidance to be taken into account during the clinical development and evaluation of currently investigated therapies for DMD and BMD [91].

As described in this chapter, major recent advances for the enablement of gene-based treatments of muscular dystrophies include the following:

- The technologies for the manipulation of the human genome have been refined to levels that achieve the high precision of gene editing that would be expected for a medicinal use. The CRISPR/Cas9-based technique has undoubtedly revolutionised the field that was pioneered through sequence-specific DNA recognition proteins based on zinc finger motifs or TALENS. It is clear that the technology now exists to make desired changes to a genome and therefore in theory to correct genetic defects. However, the remaining molecular challenges to address include (i) achieving sufficient efficiency of the modification for therapeutic benefit, (ii) controlling and/or characterising unwanted off-target genetic modifications to acceptable levels, and (iii) delivering the gene correction throughout the affected tissues of the body.

- Much of the recent renaissance of optimism in the potential of gene therapy is driven by a new emphasis on *ex vivo* treatments, especially for autologous cells. Vectors, such as lentivirus, which generally yield rather low efficiency of genetic transduction *in vivo*, become therapeutically viable when used *ex vivo* on cell populations that can also be selected and expanded before readministration to the patient. Cell production technologies have also been adapted to a paradigm based on GMP-compliant enclosed system manufacture of multiple single-patient batches in place of large-scale multipatient batches. However, although this strategy is well suited to cell types such as circulating lymphocytes (for example, for CAR-T therapies), the applicability to differentiated muscle cells is much less clear. The solution may lie in the further development of treatments using stem cells that have potential to at least partially repopulate muscle tissues with corrected cells. Stem cell-based products do have their own safety concerns to be considered, although recent experience, notably the EU approval of the limbal stem cell-based product Holoclar®, indicates that these can be addressed.

- Developers of gene-based therapies have also long been concerned that the standards required by medicine regulatory authorities would be very difficult to meet with these complex biological products, but the EMA has demonstrated that its standards are pragmatically science-based and achievable. At the time of writing, the gene therapy and oncolytic viral product Imlygic® has been recommended for approval to add to the previously approved Glybera®. Several autologous cell-based tissue engineered therapeutics have been approved, namely, ChondroCelect®, MACI®, and Holoclar®, and *ex vivo* genetically modified cell-based therapies are either under review or being readied for submission. It can therefore be concluded that there is no fundamental obstacle to approval of either *in vivo* or *ex vivo* gene therapies.

Author details

Houria Bachtarzi* and Tim Farries

*Address all correspondence to: houria.bachtarzi@eraconsulting.com

ERA Consulting (UK) Ltd. (European Regulatory Affairs), London Gas Museum, Twelvetrees Crescent, London, E3 3JG, United Kingdom

References

[1] Blake D.J., et al., Function and genetics of dystrophin and dystrophin-related proteins in muscle. Physiol Rev, 2002. 82: pp. 291–329.

[2] Mercuri E. and Muntoni F., Muscular dystrophies. Lancet, 2013. 381: pp. 845–860.

[3] Theadom A., et al., Prevalence of muscular dystrophies: a systematic literature review. Neuroepidemiology, 2014. 43: pp. 259–268.

[4] Liew W.K. and Kang P.B., Recent developments in the treatment of Duchenne muscular dystrophy and spinal muscular atrophy. Ther Adv Neurol Disord, 2013. 6: pp. 147–160.

[5] Ibraghimov-Beskrovnaya O., et al., Primary structure of dystrophin-associated glycoproteins linking dystrophin to the extracellular matrix. Nature, 1992. 355: pp. 696–702.

[6] Bushby K., et al., Diagnosis and management of Duchenne muscular dystrophy, part 1: diagnosis, and pharmacological and psychosocial management. Lancet Neurol, 2010. 9: pp. 77–93.

[7] Millay D.P., et al., Genetic and pharmacologic inhibition of mitochondrial-dependent necrosis attenuates muscular dystrophy. Nat Med, 2008. 14: pp. 442–447.

[8] Cirak S., et al., Restoration of the dystrophin-associated glycoprotein complex after exon skipping therapy in Duchenne muscular dystrophy. Mol Ther, 2012. 20: pp. 462–467.

[9] Romero N.B., et al., Phase I study of dystrophin plasmid-based gene therapy in Duchenne/Becker muscular dystrophy. Hum Gene Ther, 2004. 15: pp. 1065–1076.

[10] Mendell J.R., et al., Dystrophin immunity in Duchenne's muscular dystrophy. N Engl J Med, 2010. 363: pp. 1429–1437.

[11] Odom G.L., et al., Microutrophin delivery through rAAV6 increases lifespan and improves muscle function in dystrophic dystrophin/utrophin-deficient mice. Mol Ther, 2008. 16: pp. 1539–1545.

[12] Koo T., et al., Triple trans-splicing adeno-associated virus vectors capable of transferring the coding sequence for full-length dystrophin protein into dystrophic mice. Hum Gene Ther, 2014. 25: pp. 98–108.

[13] Lostal W., et al., Full-length dystrophin reconstitution with adeno-associated viral vectors. Hum Gene Ther, 2014. 25: pp. 552–562.

[14] Koenig M., et al., The molecular basis for Duchenne versus Becker muscular dystrophy: correlation of severity with type of deletion. Am J Hum Genet, 1989. 45: pp. 498–506.

[15] Dunckley M.G., et al., Modification of splicing in the dystrophin gene in cultured Mdx muscle cells by antisense oligoribonucleotides. Hum Mol Genet, 1998. 7: pp. 1083–1090.

[16] van Deutekom J.C., et al., Antisense-induced exon skipping restores dystrophin expression in DMD patient derived muscle cells. Hum Mol Genet, 2001. 10: pp. 1547–1554.

[17] Cirak S., et al., Exon skipping and dystrophin restoration in patients with Duchenne muscular dystrophy after systemic phosphorodiamidate morpholino oligomer treatment: an open-label, phase 2, dose-escalation study. Lancet, 2011. 378: pp. 595–605.

[18] Braun S., Gene-based therapies of neuromuscular disorders: an update and the pivotal role of patient organizations in their discovery and implementation. J Gene Med, 2013. 15: pp. 397–413.

[19] Corrigan-Curay J., et al., Genome editing technologies: defining a path to clinic. Mol Ther, 2015. 23: pp. 796–806.

[20] Smith J., et al., Requirements for double-strand cleavage by chimeric restriction enzymes with zinc finger DNA-recognition domains. Nucleic Acids Res, 2000. 28: pp. 3361–3369.

[21] Handel E.M., et al., Expanding or restricting the target site repertoire of zinc-finger nucleases: the inter-domain linker as a major determinant of target site selectivity. Mol Ther, 2009. 17: pp. 104–111.

[22] Miller J.C., et al., An improved zinc-finger nuclease architecture for highly specific genome editing. Nat Biotechnol, 2007. 25: pp. 778–785.

[23] Doyon Y., et al., Enhancing zinc-finger-nuclease activity with improved obligate heterodimeric architectures. Nat Methods, 2011. 8: pp. 74–79.

[24] Szczepek M., et al., Structure-based redesign of the dimerization interface reduces the toxicity of zinc-finger nucleases. Nat Biotechnol, 2007. 25: pp. 786–793.

[25] Sollu C., et al., Autonomous zinc-finger nuclease pairs for targeted chromosomal deletion. Nucleic Acids Res, 2010. 38: pp. 8269–8276.

[26] Guo J., et al., Directed evolution of an enhanced and highly efficient *Fok*I cleavage domain for zinc finger nucleases. J Mol Biol, 2010. 400: pp. 96–107.

[27] Sebastiano V., et al., *In situ* genetic correction of the sickle cell anemia mutation in human induced pluripotent stem cells using engineered zinc finger nucleases. Stem Cells, 2011. 29: pp. 1717–1726.

[28] Anguela X.M., et al., Robust ZFN-mediated genome editing in adult hemophilic mice. Blood, 2013. 122: pp. 3283–3287.

[29] Yusa K., et al., Targeted gene correction of α1-antitrypsin deficiency in induced pluripotent stem cells. Nature, 2011. 478: pp. 391–394.

[30] Urnov F.D., et al., Highly efficient endogenous human gene correction using designed zinc-finger nucleases. Nature, 2005. 435: pp. 646–651.

[31] Perez E.E., et al., Establishment of HIV-1 resistance in $CD4^+$ T cells by genome editing using zinc-finger nucleases. Nat Biotechnol, 2008. 26: pp. 808–816.

[32] Tebas P., et al., Gene editing of CCR5 in autologous CD4 T cells of persons infected with HIV. N Engl J Med, 2014. 370: pp. 901–910.

[33] Ousterout D.G., et al., Correction of dystrophin expression in cells from duchenne muscular dystrophy patients through genomic excision of exon 51 by zinc finger nucleases. Mol Ther, 2015. 23: pp. 523–532.

[34] Boch J. and Bonas U., *Xanthomonas* AvrBs3 family-type III effectors: discovery and function. Annu Rev Phytopathol, 2010. 48: pp. 419–436.

[35] Boch J., et al., Breaking the code of DNA binding specificity of TAL-type III effectors. Science, 2009. 326: pp. 1509–1512.

[36] Mahfouz M.M., et al., De novo-engineered transcription activator-like effector (TALE) hybrid nuclease with novel DNA binding specificity creates double-strand breaks. Proc Natl Acad Sci U S A, 2011. 108: pp. 2623–2628.

[37] Ousterout D.G., et al., Reading frame correction by targeted genome editing restores dystrophin expression in cells from Duchenne muscular dystrophy patients. Mol Ther, 2013. 21: pp. 1718–1726.

[38] Grobet L., et al., A deletion in the bovine myostatin gene causes the double-muscled phenotype in cattle. Nat Genet, 1997. 17: pp. 71–74.

[39] Schuelke M., et al., Myostatin mutation associated with gross muscle hypertrophy in a child. N Engl J Med, 2004. 350: pp. 2682–2688.

[40] Mosher D.S., et al., A mutation in the myostatin gene increases muscle mass and enhances racing performance in heterozygote dogs. PLoS Genet, 2007. 3: p. e79.

[41] Bogdanovich S., et al., Functional improvement of dystrophic muscle by myostatin blockade. Nature, 2002. 420: pp. 418–421.

[42] Roth S.M. and Walsh S., Myostatin: a therapeutic target for skeletal muscle wasting. Curr Opin Clin Nutr Metab Care, 2004. 7: pp. 259–263.

http://dx.doi.org/10.5772/62282

[43] Sakuma K. and Yamaguchi A., Molecular mechanisms in aging and current strategies to counteract sarcopenia. Curr Aging Sci, 2010. 3: pp. 90–101.

[44] Bishop A., et al., The therapeutic potential of agents that inactivate myostatin. Expert Opin Investig Drugs, 2005. 14: pp. 1099–1106.

[45] Lee S.J. and Glass D.J., Treating cancer cachexia to treat cancer. Skelet Muscle, 2011. 1: pp. 2.

[46] Xu L., et al., Targeted myostatin gene editing in multiple mammalian species directed by a single pair of TALE nucleases. Mol Ther Nucleic Acids, 2013. 2: p. e112.

[47] Wiedenheft B., et al., RNA-guided genetic silencing systems in bacteria and archaea. Nature, 2012. 482: pp. 331–338.

[48] Sontheimer E.J. and Barrangou R., The bacterial origins of the CRISPR genome-editing revolution. Hum Gene Ther, 2015. 26: pp. 413–424.

[49] Koonin E.V. and Makarova K.S., CRISPR-Cas: evolution of an RNA-based adaptive immunity system in prokaryotes. RNA Biol, 2013. 10: pp. 679–686.

[50] Makarova K.S., et al., Evolution and classification of the CRISPR-Cas systems. Nat Rev Microbiol, 2011. 9: pp. 467–477.

[51] Mali P., et al., RNA-guided human genome engineering via Cas9. Science, 2013. 339: pp. 823–826.

[52] Long C., et al., Prevention of muscular dystrophy in mice by CRISPR/Cas9-mediated editing of germline DNA. Science, 2014. 345: pp. 1184–1188.

[53] Davie A.M. and Emery A.E., Estimation of proportion of new mutants among cases of Duchenne muscular dystrophy. J Med Genet, 1978. 15: pp. 339–345.

[54] Ousterout D.G., et al., Multiplex CRISPR/Cas9-based genome editing for correction of dystrophin mutations that cause Duchenne muscular dystrophy. Nat Commun, 2015. 6: pp. 6244.

[55] Canver M.C., et al., Characterization of genomic deletion efficiency mediated by clustered regularly interspaced palindromic repeats (CRISPR)/Cas9 nuclease system in mammalian cells. J Biol Chem, 2014. 289: pp. 21312–21324.

[56] Nathwani A.C., et al., Long-term safety and efficacy following systemic administration of a self-complementary AAV vector encoding human FIX pseudotyped with serotype 5 and 8 capsid proteins. Mol Ther, 2011. 19: pp. 876–885.

[57] Wang Z., et al., Adeno-associated virus serotype 8 efficiently delivers genes to muscle and heart. Nat Biotechnol, 2005. 23: pp. 321–328.

[58] Inagaki K., et al., Robust systemic transduction with AAV9 vectors in mice: efficient global cardiac gene transfer superior to that of AAV8. Mol Ther, 2006. 14: pp. 45–53.

[59] Long C., et al., Postnatal genome editing partially restores dystrophin expression in a mouse model of muscular dystrophy. Science, 2015.

[60] Nelson C.E., et al., *In vivo* genome editing improves muscle function in a mouse model of Duchenne muscular dystrophy. Science, 2015.

[61] Tabebordbar M., et al., *In vivo* gene editing in dystrophic mouse muscle and muscle stem cells. Science, 2015.

[62] Harish P., et al., Progress on gene therapy, cell therapy, and pharmacological strategies toward the treatment of oculopharyngeal muscular dystrophy. Hum Gene Ther, 2015. 26: pp. 286–292.

[63] Verheesen P., et al., Prevention of oculopharyngeal muscular dystrophy-associated aggregation of nuclear polyA-binding protein with a single-domain intracellular antibody. Hum Mol Genet, 2006. 15: pp. 105–111.

[64] Bachtarzi H., et al., Adeno-associated virus (AAV)-mediated RNA interference to PABPN1 combined with an optimised resistant transgene for rescue of the muscle specific disease oculopharyngeal muscular dystrophy (OPMD). Abstract from 7th Annual UK Neuromuscular Translational Research Conference 2014, London, United Kingdom. Neuromuscular Disorders, 2014.

[65] Turner C. and Hilton-Jones D., The myotonic dystrophies: diagnosis and management. J Neurol Neurosurg Psychiatry, 2010. 81: pp. 358–367.

[66] Gharehbaghi-Schnell E.B., et al., Genotype-phenotype correlation in myotonic dystrophy. Clin Genet, 1998. 53: pp. 20–26.

[67] Hamshere M.G., et al., Myotonic dystrophy: the correlation of (CTG) repeat length in leucocytes with age at onset is significant only for patients with small expansions. J Med Genet, 1999. 36: pp. 59–61.

[68] Furling D., et al., Insulin-like growth factor I circumvents defective insulin action in human myotonic dystrophy skeletal muscle cells. Endocrinology, 1999. 140: pp. 4244–4250.

[69] Trip J., et al., Drug treatment for myotonia. Cochrane Database Syst Rev, 2006: p. CD004762.

[70] Furling D., et al., Viral vector producing antisense RNA restores myotonic dystrophy myoblast functions. Gene Ther, 2003. 10: pp. 795–802.

[71] Sobczak K., et al., RNA interference targeting CUG repeats in a mouse model of myotonic dystrophy. Mol Ther, 2013. 21: pp. 380–387.

[72] Bisset D.R., et al., Therapeutic impact of systemic AAV-mediated RNA interference in a mouse model of myotonic dystrophy. Hum Mol Genet, 2015. 24: pp. 4971–4983.

[73] Xia G., et al., Genome modification leads to phenotype reversal in human myotonic dystrophy type 1 induced pluripotent stem cell-derived neural stem cells. Stem Cells, 2015. 33: pp. 1829–1838.

[74] Gabriel R., et al., An unbiased genome-wide analysis of zinc-finger nuclease specificity. Nat Biotechnol, 2011. 29: pp. 816–823.

[75] Suzuki K., et al., Targeted gene correction minimally impacts whole-genome mutational load in human-disease-specific induced pluripotent stem cell clones. Cell Stem Cell, 2014. 15: pp. 31–36.

[76] Sadelain M., et al., Safe harbours for the integration of new DNA in the human genome. Nat Rev Cancer, 2012. 12: pp. 51–58.

[77] Corrigan-Curay J., et al., Challenges in vector and trial design using retroviral vectors for long-term gene correction in hematopoietic stem cell gene therapy. Mol Ther, 2012. 20: pp. 1084–1094.

[78] Porter S.N., et al., Lentiviral and targeted cellular barcoding reveals ongoing clonal dynamics of cell lines *in vitro* and *in vivo*. Genome Biol, 2014. 15: pp. R75.

[79] Mussolino C., et al., TALENs facilitate targeted genome editing in human cells with high specificity and low cytotoxicity. Nucleic Acids Res, 2014. 42: pp. 6762–6773.

[80] Yla-Herttuala S., Endgame: glybera finally recommended for approval as the first gene therapy drug in the European Union. Mol Ther, 2012. 20: pp. 1831–1832.

[81] Chen X. and Goncalves M.A., Engineered viruses as genome editing devices. Mol Ther, 2015.

[82] EMA, Reflection paper on quality, non-clinical and clinical issues related to the development of recombinant adeno-associated viral vectors. 2010. EMEA/CHMP/GTWP/587488/2007 Rev. 1.

[83] Bachrach E., et al., Systemic delivery of human microdystrophin to regenerating mouse dystrophic muscle by muscle progenitor cells. Proc Natl Acad Sci U S A, 2004. 101: pp. 3581–3586.

[84] Li S., et al., Stable transduction of myogenic cells with lentiviral vectors expressing a minidystrophin. Gene Ther, 2005. 12: pp. 1099–1108.

[85] Kobinger G.P., et al., Correction of the dystrophic phenotype by *in vivo* targeting of muscle progenitor cells. Hum Gene Ther, 2003. 14: pp. 1441–1449.

[86] Perie S., et al., Autologous myoblast transplantation for oculopharyngeal muscular dystrophy: a phase I/IIa clinical study. Mol Ther, 2014. 22: pp. 219–225.

[87] Tedesco F.S., et al., Transplantation of genetically corrected human iPSC-derived progenitors in mice with limb-girdle muscular dystrophy. Sci Transl Med, 2012. 4: p. 140ra89.

[88] Ikemoto M., et al., Autologous transplantation of SM/C-2.6(+) satellite cells transduced with micro-dystrophin CS1 cDNA by lentiviral vector into mdx mice. Mol Ther, 2007. 15: pp. 2178–2185.

[89] Quenneville S.P., et al., Autologous transplantation of muscle precursor cells modified with a lentivirus for muscular dystrophy: human cells and primate models. Mol Ther, 2007. 15: pp. 431–438.

[90] EMA, Guideline on the quality, non-clinical and clinical aspects of gene therapy medicinal products. 2015. EMA/CAT/80183/2014.

[91] EMA, Guideline on the clinical investigation of medicinal products for the treatment of Duchenne and Becker muscular dystrophy. 2015. EMA/CHMP/236981/2011, Corr. 1.